# MICRO IRRIGATION ENGINEERING FOR HORTICULTURAL CROPS

## Policy Options, Scheduling, and Design

# MICRO IRRIGATION ENGINEERING FOR HORTICULTURAL CROPS

## Policy Options, Scheduling, and Design

*Edited by*
**Ajai Singh, PhD, FIE**
**Megh R. Goyal, PhD, PE**

**AAP** APPLE ACADEMIC PRESS

Apple Academic Press Inc.          Apple Academic Press Inc.
3333 Mistwell Crescent             9 Spinnaker Way
Oakville, ON L6L 0A2 Canada        Waretown, NJ 08758 USA

© 2017 by Apple Academic Press, Inc.

First issued in paperback 2021

*Exclusive worldwide distribution by CRC Press, a member of Taylor & Francis Group*
No claim to original U.S. Government works

ISBN 13: 978-1-77-463664-0 (pbk)
ISBN 13: 978-1-77-188540-9 (hbk)

---

**Library and Archives Canada Cataloguing in Publication**

---

Micro irrigation engineering for horticultural crops : policy options, scheduling, and design / edited by Ajai Singh, PhD, FIE, Megh R. Goyal, PhD, PE.

(Innovations and challenges in micro irrigation)
Includes bibliographical references and index.
Issued in print and electronic formats.
ISBN 978-1-77188-540-9 (hardcover).--ISBN 978-1-315-20742-1 (PDF)

1. Microirrigation. I. Goyal, Megh Raj,editor II. Singh, Ajai, 1970-, editor III. Series: Innovations and challenges in micro irrigation

S619 T74 M53 2017          631.5'87          C2017-903112-0          C2017-903113-9

..................................................................................................................................................

CIP data on file with US Library of Congress

..................................................................................................................................................

Apple Academic Press also publishes its books in a variety of electronic formats. Some content that appears in print may not be available in electronic format. For information about Apple Academic Press products, visit our website at **www.appleacademicpress.com** and the CRC Press website at **www.crcpress.com**

# CONTENTS

# LIST OF CONTRIBUTORS

**Vaibhav Bhamoriya, PhD**
Assistant Professor in Economics, Indian Institute of Management Kashipur, Bazaar Road, Kashipur District, Udham Singh Nagar 244713, Uttarakhand, India. E-mail: vaibhavb@iima.ac.in; vaibhavb@iimahd.ernet.in

**H. S. Chauhan, PhD**
Former Professor, Irrigation and Drainage Engineering Department, College of Technology, G.B. Pant University of Agriculture and Technology, Pantnagar 263145, Uttarakhand, India.

**Murari Lal Gaur, PhD (Hydrology)**
Professor and Head, Soil and Water Engineering Department; and Principal at College of Agricultural Engineering and Technology, Anand Agricultural University, Godhra 389001, Gujarat, India. E-mail: mlgaur@yahoo.com; dr.mlgaur@gmail.com

**Megh R. Goyal, PhD, PE**
Retired Professor in Agricultural and Biomedical Engineering, University of Puerto Rico, Mayaguez Campus; and Senior Technical Editor-in-Chief in Agriculture Sciences and Biomedical Engineering, Apple Academic Press Inc., PO Box 86, Rincon, PR 00677, USA. E-mail: goyalmegh@gmail.com

**Murtaza Hasan, PhD**
Centre for Protected Cultivation and Technology, Indian Agricultural Research Institute (ICAR), PUSA, New Delhi 110012, India. E-mail: mhasan_indo@iari.res.in

**Ashwani Kumar Madile, MTech**
Department of Irrigation and Drainage Engineering, G.B. Pant University of Agriculture and Technology, Pantnagar, Udham Singh Nagar 263145, Uttarakhand, India. E-mail: a10madile@gmail.com

**M. V. Manjunatha, PhD**
Department of Agricultural Engineering, University of Agricultural Sciences, Dharwad 580005, Karnataka, India. E-mail: mvmuasd@gmail.com

**T. Mohanasundari, PhD**
Tamil Nadu Agricultural University, Coimbatore 641003, Tamil Nadu, India.

**K. Palanisami, PhD**
International Water Management Institute (IWMI), ICRISAT Campus, 401/5, Patancheru, Medak 502324, Telangana, India. E-mail: k.palanisami@cgiar.org

**S. Raman, PhD**
Water Resources Expert and Consultant, Mumbai, India.

**K. Krishna Reddy, PhD**
International Water Management Institute (IWMI), ICRISAT Campus, 401/5, Patancheru, Medak 502324, Telangana, India. E-mail: iwmi-hyderabad@cgiar.org

**Ajai Singh, PhD, FIE**
Associate Professor and Head, Centre for Water Engineering and Management, Central University of Jharkhand, Brambe 834205, Ranchi, India. E-mail: ajai_jpo@yahoo.com; ajai.singh@cuj.ac.in

**K. K. Singh, PhD**
Irrigation and Drainage Engineering Department, College of Technology, G. B. Pant University of Agriculture and Technology, Pantnagar, Udham Singh Nagar 263145, Uttarakhand, India.

**P. K. Singh, PhD**
Irrigation and Drainage Engineering Department, College of Technology, G. B. Pant University of Agriculture and Technology, Pantnagar, Udham Singh Nagar 263145, Uttarakhand, India. E-mail: singhpk67@gmail.com

**R. Singh, MTech**
Irrigation and Drainage Engineering Department, College of Technology, G. B. Pant University of Agriculture and Technology, Pantnagar, Udham Singh Nagar 263145, Uttarakhand, India.

**Surjeet Singh, PhD**
National Institute of Hydrology, Roorkee, Uttarakhand, India.

**S. K. Srivastava, PhD**
Vauge School of Agricultural Engineering, SHIATS, Allahabad 211007, India. E-mail: santoshagri.2008@rediffmail.com

# LIST OF ABBREVIATIONS

| | |
|---|---|
| AP | Andhra Pradesh |
| APMIP | Andhra Pradesh Micro Irrigation Project |
| ASAE | American Society of Agricultural Engineers |
| bcm | billion cubic meters |
| BCR | benefit cost ratio |
| CPE | cumulative pan evaporation |
| CRF | capital recovery factor |
| CSS | central sponsored scheme |
| CSWI | canopy water stress index |
| CWP | crop water productivity |
| CWR | crop water requirement |
| DBTL | direct benefit transfer for loan |
| DDP | Desert Development Program |
| DI | drip irrigation |
| DPAP | Drought-Prone Area Program |
| EC | electrical conductivity |
| ET | evapotranspiration |
| FAO | Food and Agriculture Organization |
| FUE | fertilizer use efficiency |
| GH | greenhouse |
| GGRC | Gujarat Green Revolution Company Ltd. |
| GOI | Government of India |
| HDPE | high-density polyethylene |
| HSPA | Hawaiian Sugar Planter's Association |
| ICU | irrigation control unit |
| INCID | Indian Committee on Irrigation and Drainage |
| INR | Indian Rupees |
| IPE | irrigation production efficiency |
| IRR | internal rate of return |
| IWMI | International Water Management Institute |
| LDPE | low-density polyethylene |
| MI | micro irrigation |

| | |
|---|---|
| NCPAH | National Committee on Plasticulture Application in Horticulture |
| NMMI | National Mission on Micro Irrigation |
| NMSA | National Mission on Sustainable Agriculture |
| NUE | nutrient use efficiency |
| NWP | nutritional water productivity |
| OBC | other backward classes |
| OFWM | on-farm water management |
| PMKSY | Pradhan Manthri Krishi Sinchayee Yojana |
| PVC | polyvinyl chloride |
| SC | scheduled caste |
| SDI | subsurface drip irrigation |
| SMP | soil moisture potential |
| SPV | special-purpose vehicle |
| SRI | System of Rice Intensification |
| TN | Tamil Nadu |
| UCH | Hart uniformity coefficient |
| WUE | water use efficiency |

# FOREWORD

This book, under the book series "Innovations and Challenges in Micro Irrigation," encompasses the relevant research work on micro irrigation and can be quite useful for graduate students and practicing engineers. We need to focus on innovation and evolving new paradigms for efficient utilization of water resources as a means of socioeconomic development of humankind. Water is an essential natural resource for life-supporting systems of all living beings. It is the single most important input in agriculture and has a major role in providing stability and enhancement of agricultural production, leading to self-sufficiency and sustainability. Therefore, application of micro irrigation systems can play an important role to achieve the aim of sustainable development and healthy ecosystems. The per capita availability of water is dwindling and approaching the scarcity levels not far in the future. There is immense need to conserve and use most efficiently both surface water and groundwater resources.

Prof. Megh R. Goyal, Senior Editor-in-Chief of 20 books on micro irrigation by Apple Academic Press Inc. (AAP) and Father of Irrigation Engineering of 21st Century in Puerto Rico, has edited this book volume. I am happy to learn that Dr. Ajai Singh of Central University of Jharkhand, Ranchi, India, has joined him, and both the editors have made commendable efforts to bring this book volume. I also like to commend efforts by AAP to publish quality books on micro irrigation.

I wish the authors all the success in this as well as in future endeavor in this direction.

**Nand Kumar Yadav "Indu," PhD**
Vice Chancellor and Professor
Central University of Jharkhand
Brambe, Ranchi 834205, India.

# PREFACE 1

Adoption of micro irrigation systems can be a panacea in irrigation-related problems and will increase the area under cultivation. In this technology, the cropped field is irrigated in the close vicinity of root zone of crop. It reduces the water loss occurring through evaporation, conveyance, and distribution. Therefore, high water use efficiency can be achieved. The rain-fed cropped area can be increased with this technology, and potential sources of food production for the benefit of world's food security could be augmented. This edited book has chapters ranging from policy intervention to application of systems to different crops and even under different land conditions. This has been a continued effort of Prof. Goyal to compile the research works in a form of a book series and provide an opportunity for the large scientific community to have easy access.

I feel very fortunate to work with Dr. Megh R. Goyal, who indeed made a serious effort to invite quality chapters. I owe my deepest gratitude to Prof. Nand Kumar Yadav "Indu," Vice Chancellor at Central University of Jharkhand for his support and encouragement. The editors are grateful to many individuals who have contributed their works in the form of chapters.

I feel profound privilege in expressing my heartfelt reverence to my parents, brothers and sister, in-laws for their blessings and moral support to achieve this goal. Last but not the least, I acknowledge with heartfelt indebtedness, the patience and the generous support rendered by my wife, Punam, and our daughter, Anushka, who always allowed me to work continuously and relentlessly.

**—Ajai Singh, PhD, FIE**

# PREFACE 2

During October 22 through November 4, 2015, I along with my wife visited UNESCO World Heritage archeological sites in Athens (Ἀθῆναι *Athēnai*), Corinthia (Greek: Κορινθία-*Korinthía*), Ephesus, Malta, and Rome.

My vision for micro irrigation technology has expanded globally. I am surprised to observe how this is expanding to tourist regions and especially to archeological sites with number of visitors exceeding 1 million per year. Although no emphasis is made to draw attention of visitors to this valuable technology, yet there is a potential audience. At one of these sites, I started my own initiative to explain this water-saving technique to a small group along with the administrator of this site, who happened to be a civil engineer. He promised me to promote this through a short presentation, of course at a nominal cost.

Water being the limited resource, its efficient use is essential in order to increase agricultural production per unit volume of water and per unit area of crop land. Due to increase in the population, the competition of limited water resources for domestic, industrial, and agricultural needs is increasing considerably. Water for irrigation is becoming scarce and expensive due to depletion of surface and subsurface water caused by erratic rainfall and overexploitation. Therefore, it is essential to formulate economically viable water and other input management strategies in order to irrigate more land area with existing water resources and to enhance crop productivity. Improper distribution lowers the conveyance efficiency and ultimately causes water loss. Therefore, right amount at right time and frequency of irrigation is vital for optimum use of limited water resources for crop production and management.

The aim of irrigation scheduling is to increase efficiencies by applying the exact amount of water needed to replenish the soil moisture to the desired level. Appropriate irrigation scheduling saves water and energy. Therefore, it is important to develop irrigation scheduling techniques under prevailing climatic conditions in order to utilize scarce water resources effectively for crop production. Numerous studies have been carried out in the past in the development and evaluation of irrigation scheduling under a wide range of irrigation systems and management, soil, crop, and agroclimatic conditions. Climate-based irrigation scheduling approaches (such as pan evaporation replenishment and cumulative pan evaporation and ratio of irrigation water to cumulative pan evaporation) have been used by many researchers due to simplicity, data availability, and higher degree of adaptability at the farmer's field. Surface irrigation is the most common method for field/vegetable/fruit crops and ornamental plants. The overall efficiency of surface irrigation method is considerably low compared to modern pressurized irrigation systems: drip, micro-sprinkler, and sprinkler. Drip irrigation can potentially provide high application efficiency and application uniformity.

This book volume presents policy adoption methods, irrigation scheduling, and design procedures in micro irrigation engineering for horticultural crops.

The mission of this book volume is to serve as a reference manual for graduate and undergraduate students of agricultural, biological, and civil engineering; horticulture, soil science, crop science, and agronomy. I hope that it will be a valuable reference for professionals that work with micro irrigation and water management; for professional training institutes, technical agricultural centers, irrigation centers, agricultural extension services, and other agencies that work with micro irrigation programs.

After my first textbook, *Drip/Trickle or Micro Irrigation Management* by Apple Academic Press Inc., and response from international readers, Apple Academic Press Inc. has published for the world community the 10-volume series on *Research Advances in Sustainable Micro Irrigation* edited by M. R. Goyal. The website <appleacademicpress.com> gives details on these 10 book volumes.

This book is volume six of the book series *Innovations and Challenges in Micro Irrigation*. Both books series are a must for those interested in irrigation planning and management, namely, researchers, scientists, educators, and students.

The contributions by the cooperating authors to this book series have been most valuable in the compilation of this volume. Their names are mentioned in each chapter and in the list of contributors. This book would not have been written without the valuable cooperation of Dr. Ajai Singh and the investigators, many of whom are renowned scientists who have worked in the field of micro irrigation throughout their professional careers.

I would like to thank editorial staff, Sandy Jones Sickels, Vice President, and Ashish Kumar, Publisher and President at Apple Academic Press, Inc., for making every effort to publish the book when the diminishing water resources are a major issue worldwide. Special thanks are due to the AAP production staff for the quality production of this book.

We request the reader to offer us your constructive suggestions that may help to improve the next edition.

I express my deep admiration to my wife, Subhadra Devi Goyal, for understanding and collaboration during the preparation of this book. I dedicate this book volume to research scientists at the Water Technology Centre of Tamil Nadu Agricultural University, who made earnest efforts to water conservation practices in Southern India.

As an educator, there is a piece of advice to one and all in the world: "Permit that our almighty God, our Creator, excellent Teacher and Micro Irrigation Designer, irrigate our life with His Grace of rain trickle by trickle, because our life must continue trickling on...."

**—Megh R. Goyal, PhD, PE**
Senior Editor-in-Chief

# WARNING/DISCLAIMER

**PLEASE READ CAREFULLY**

The goal of this compendium, *Micro Irrigation Engineering for Horticultural Crops,* is to guide the world engineering community on how to efficiently employ micro irrigation engineering for horticultural agriculture. The reader must be aware that the dedication, commitment, honesty, and sincerity are most important factors in a dynamic manner for a complete success.

The editors, the contributing authors, the publisher and the printer have made every effort to make this book as complete and as accurate as possible. However, there still may be grammatical errors or mistakes in the content or typography. Therefore, the contents in this book should be considered as a general guide and not a complete solution to address any specific situation in irrigation. For example, fruit or vegetable or meat or grain, etc. requires a different type of engineering intervention to process such produce.

The editors, the contributing authors, the publisher and the printer shall have neither liability nor responsibility to any person, any organization or entity with respect to any loss or damage caused, or alleged to have caused, directly or indirectly, by information or advice contained in this book. Therefore, the purchaser/reader must assume full responsibility for the use of the book or the information therein.

The mention of commercial brands and trade names are only for technical purposes. We do not endorse particular products or equipment mentioned.

All web-links that are mentioned in this book were active on December 31, 2016. The editors, the contributing authors, the publisher and the printing company shall have neither liability nor responsibility, if any of the web-links is inactive at the time of reading of this book.

# ABOUT THE LEAD EDITOR

 **Ajai Singh, PhD,** FIE, is an Associate Professor and Head of the Centre for Water Engineering and Management at the Central University of Jharkhand in Ranchi, India. He is a Fellow of the Institution of Engineers (India) and a life member of the Indian Society of Agricultural Engineers, the Indian Water Resources Society, the Indian Association of Hydrologist, the Indian Meteorological Society, and the Crop and Weed Science Society. Formerly he has worked as junior project officer at the Indian Institute of Technology (IIT) Kharagpur; and then as Assistant Professor at North Bengal Agricultural University, West Bengal.

Dr. Singh has authored more than 30 articles in technical journals and textbooks, including the book *Introduction to Drip Irrigation*. He has also written the books *Hydrological Modelling Using Process Based and Data Driven Models* and *Finite Element Analysis and Optimal Design of Drip Irrigation Sub-main*. His area of active research is hydrological modeling, micro irrigation engineering, water resources planning and management, and groundwater hydrology. He has been conferred the Distinguished Services Certificate (2012) by Indian Society of Agricultural Engineers, New Delhi.

His area of active research is hydrological modeling, micro irrigation engineering, water resources planning and management, and groundwater hydrology. He has been conferred Distinguished Services Certificate (2012) by Indian Society of Agricultural Engineers, New Delhi.

Dr. Singh received a BTech degree in Agricultural Engineering in 1995 and completed his MTech. in 1997. He obtained his PhD degree in 2011.

Readers may contact him at: ajai.singh@cuj.ac.in

# ABOUT THE SENIOR EDITOR-IN-CHIEF

Megh R. Goyal, PhD, PE, is, at present, a Retired Professor in Agricultural and Biomedical Engineering from the General Engineering Department in the College of Engineering at University of Puerto Rico, Mayaguez Campus; and Senior Acquisitions Editor and Senior Technical Editor-in-Chief in Agricultural and Biomedical Engineering for Apple Academic Press Inc.

He received his BSc degree in Engineering in 1971 from Punjab Agricultural University, Ludhiana, India; his MSc degree in 1977; and PhD degree in 1979 from the Ohio State University, Columbus; his Master of Divinity degree in 2001 from Puerto Rico Evangelical Seminary, Hato Rey, Puerto Rico, USA.

Since 1971, he has worked as Soil Conservation Inspector (1971); Research Assistant at Haryana Agricultural University (1972–1975) and the Ohio State University (1975–1979); Research Agricultural Engineer/ Professor at Department of Agricultural Engineering of UPRM (1979– 1997); and Professor in Agricultural and Biomedical Engineering at General Engineering Department of UPRM (1997–2012). He spent 1-year sabbatical leave in 2002–2003 at Biomedical Engineering Department, Florida International University, Miami, USA.

He was the first agricultural engineer to receive the professional license in Agricultural Engineering in 1986 from College of Engineers and Surveyors of Puerto Rico. On September 16, 2005, he was proclaimed as "Father of Irrigation Engineering in Puerto Rico for the Twentieth Century" by the ASABE, Puerto Rico Section, for his pioneer work on micro irrigation, evapotranspiration, agroclimatology, and soil and water engineering. During his professional career of 45 years, he has received awards such as: Scientist of the Year, Blue Ribbon Extension Award, Research Paper Award, Nolan Mitchell Young Extension Worker Award,

Agricultural Engineer of the Year, Citations by Mayors of Juana Diaz and Ponce, Membership Grand Prize for ASAE Campaign, Felix Castro Rodriguez Academic Excellence, Rashtrya Ratan Award and Bharat Excellence Award and Gold Medal, Domingo Marrero Navarro Prize, Adopted son of Moca, Irrigation Protagonist of UPRM, Man of Drip Irrigation by Mayor of Municipalities of Mayaguez/Caguas/Ponce, and Senate/Secretary of Agriculture of ELA, Puerto Rico.

Dr. Megh R. Goyal has been recognized as one of the experts "who rendered meritorious service for the development of [the] irrigation sector in India." This honor was bestowed by the Water Technology Centre of Tamil Nadu Agricultural University in Coimbatore, India, to Dr. Goyal during the inaugural session of the National Congress on "New Challenges and Advances in Sustainable Micro Irrigation" on March 1, 2017.

He has authored more than 200 journal articles and edited more than 45 books including *Elements of Agroclimatology* (Spanish) by UNISARC, Colombia; two *Bibliographies on Drip Irrigation*.

Apple Academic Press Inc. (AAP) has published his books, namely: *Management of Drip/Trickle or Micro Irrigation*, and *Evapotranspiration: Principles and Applications for Water Management*. During 2014–2016, AAP has published his 10-volume set on *Research Advances in Sustainable Micro Irrigation*. During 2016–2017, AAP will be publishing book volumes on emerging technologies/issues/challenges under two book series, *Innovations and Challenges in Micro Irrigation*, and *Innovations in Agricultural & Biological Engineering*. Readers may contact him at: <goyalmegh@gmail.com≥

# OTHER BOOKS ON MICRO IRRIGATION TECHNOLOGY BY APPLE ACADEMIC PRESS, INC.

**Management of Drip/Trickle or Micro Irrigation**
Megh R. Goyal, PhD, PE, Senior Editor-in-Chief

**Evapotranspiration: Principles and Applications for Water Management**
Megh R. Goyal, PhD, PE, and Eric W. Harmsen, Editors

**Book Series: Research Advances in Sustainable Micro Irrigation**
Senior Editor-in-Chief: Megh R. Goyal, PhD, PE

Volume 1:  Sustainable Micro Irrigation: Principles and Practices
Volume 2:  Sustainable Practices in Surface and Subsurface Micro Irrigation
Volume 3:  Sustainable Micro Irrigation Management for Trees and Vines
Volume 4:  Management, Performance, and Applications of Micro Irrigation Systems
Volume 5:  Applications of Furrow and Micro Irrigation in Arid and Semi-Arid Regions
Volume 6:  Best Management Practices for Drip Irrigated Crops
Volume 7:  Closed Circuit Micro Irrigation Design: Theory and Applications
Volume 8:  Wastewater Management for Irrigation: Principles and Practices
Volume 9:  Water and Fertigation Management in Micro Irrigation
Volume 10: Innovation in Micro Irrigation Technology

**Book Series: Innovations and Challenges in Micro Irrigation**
Senior Editor-in-Chief: Megh R. Goyal, PhD, PE

Volume 1:  Principles and Management of Clogging in Micro Irrigation
Volume 2:  Sustainable Micro Irrigation Design Systems for Agricultural Crops: Methods and Practices

# PART I
# Policy Options: Drip Irrigation Among Adopters

# CHAPTER 1

# OPINION OF ADOPTERS AND NONADOPTERS TOWARD DRIP IRRIGATION: INSIGHTS FOR MARKETING

VAIBHAV BHAMORIYA[*]

*Indian Institute of Management Kashipur, Bazaar Road, Kashipur District, Udham Singh Nagar 244713, Uttarakhand, India.*

[*]*E-mail: vaibhavb@iima.ac.in; vaibhavb@iimahd.ernet.in*

## CONTENTS

## ABSTRACT

This chapter highlights some consumer perceptions and focuses on the differences between the adopters and nonadopters. The results indicate that there are many myths about drip irrigation and these usually make the nonadopters have very high expectations from the technology and thereby suppress the satisfaction levels post-adoption. Marketing needs to take into consideration such myths and educate the prospective customers about these and also take up activities to ensure that customers are better prepared and taken care of better to deal with the actual events post-adoption. There are many other aspects of drip irrigation that marketers need to focus: drip irrigation helps in timely and adequate availability of irrigation; it is costly and difficult to master; it enhances the chances for increasing incomes and increase in water tables, and also enables agriculture with very limited water availability while effecting a reduction in the usage of water and power consumed for irrigation and provision of very good quality after-sales service.

## 1.1   INTRODUCTION

A global crisis about water and its management is significantly about availability of water for use and its characteristic of highly uneven spatial distribution. Enhancing water availability and making it amenable for use and managing the distribution are challenges of a tall order due to the dynamic nature of the resource and its usage. Agriculture accounts for majority of global freshwater withdrawals and almost all in some fast-growing economies [13]. At the global level, more than two-thirds of the blue water withdrawals are for irrigation. Irrigated agriculture represents almost a fifth of the total cultivated land but contributes more than one-third of the total food produced worldwide [13] and therefore it is of critical importance of sustenance for the human race.

In India, for example, the area irrigated with groundwater has increased to 500% since 1960. As on 2009, annual groundwater withdrawal for irrigation has been estimated as 221 billion cubic meters (bcm). The overall irrigation efficiency in India is often quite low compared to global standards. It is believed that this is mainly because the efficiency of conventional flood irrigation technique, practiced in large parts of India, is itself very low. Thus, micro irrigation (MI) technology, including drip and sprinkler irrigation, was introduced as a water-saving technology [8]. It was expected to make a contribution to conservation of the water resources in India [9]. A minimalist expectation was to save water from the quantum used in irrigation and it was expected to promote sustainable water use [2, 8]. Various field experiments have shown this technique to increase water use efficiency up to 80–90% depending on the crop and soil type [2, 11].

Drip irrigation is one of the most efficient methods of irrigation [3]. It is also seen as a promising technology in terms of its ability to support the farmer in raising incomes and reducing poverty [14]. The benefits of MI include water saving, increased yield and productivity of certain crops (especially spaced crops), labor cost savings, electricity savings, lesser pumping hours and hence easier irrigation, better crop growth and also better soil health. A lot of evidence exists claiming economic benefits on the adoption of MI. However, there exists little or sparing evidence of socioeconomic benefits from the adoption of MI.

The rapid commercialization, of agriculture enabled by MI in various regions of India and across a variety of plantations and field crops, is resulting in higher adoption rates of MI in such areas. In spite of these advantages, the spread of MI has been restricted to only a few pockets across India.

Exploring the marketing and impact of adoption of MI is crucial as different states of India such as Gujarat, Andhra Pradesh, and Rajasthan are giving a massive push to MI due to water resource conservation. The Andhra Pradesh Micro Irrigation Project (APMIP) claimed to have brought 0.166 million ha of area under MI during 2.5 years [10]. At the same time, there are pockets like Jalgaon and Nashik in Maharashtra, Narsinghpur and Maikaal in Madhya Pradesh where the market forces are leading to high adoption rates. In some pockets, the high adoption rate is observed even in the absence of government subsidies [14].

The marketers increasingly want to sell and promote drip irrigation without subsidies as they see the potential to expand the market much faster and to more users than possible with subsidies. Therefore, they are looking for insights into consumers' attitude and understanding of drip irrigation to evolve better marketing practices.

There are several such issues related to MI in India. Understanding perception of drip irrigation system, the differences among the adopters and nonadopters is an important stepping-stone toward formulating better marketing of drip irrigation. This chapter outlines the details on differences of opinions among users and nonusers across three different states and proposes certain marketing insights based on the analyses.

## 1.2   REVIEW OF LITERATURE

### 1.2.1   THE GLOBAL WATER MANAGEMENT CRISIS AND MICRO IRRIGATION

According to the UN estimates, the volume of global freshwater resources is around 35 million km³, or about 2.5% of the total volume. Around 30% of the world's freshwater is stored in the form of groundwater (shallow and deep groundwater basins up to 2000 m, soil moisture, swamp water, and permafrost). This constitutes about 97% of all freshwater that is potentially available for human use. Freshwater lakes and rivers contain an estimated 105,000 km³ or around 0.3% of the world's freshwater [11, 12, 13].

In 1989, 63% of the world's irrigated area was in Asia, compared with 64% in 1994. Also 37% of arable land of Asia was irrigated in 1994. Among Asian countries, India has the largest arable land, which is close to 39% of Asia's arable land. Irrigated agriculture represents 20% of the total cultivated land but contributes 40% of the total food produced worldwide [12, 13]. Agriculture accounts for around 70% of global freshwater withdrawals, even up to 90% in some fast-growing economies [13].

In India, for example, the area irrigated with groundwater has increased 500% since 1960. In developing and transforming nations, this "global boom" has occurred at various economic levels: subsistence farming, staple-crop production, and commercial cash crop cultivation. It has brought major socioeconomic benefits to many rural communities in Asia, the Middle East and North Africa and Latin America—with numerous countries establishing large groundwater-dependent economies [16]. The countries with the largest extent of areas equipped for irrigation with groundwater, in absolute terms, are India (39 million ha), China (19 million ha), and the USA (17 million ha). Figure 1.1 shows that 91% of the water withdrawal in the country is for irrigation and livestock purposes.

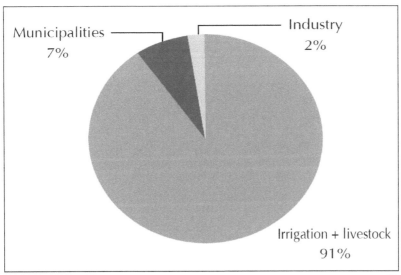

**FIGURE 1.1**    Water withdrawal in India for 2010 (From AquaStat. Country Profile—India, 2013)

Also one-third of the water withdrawal in India was from groundwater sources in 2010, which is gaining increasing prominence with each passing year. Today, groundwater supports approximately 60% of irrigated agriculture and more than 80% of rural and urban water supplies in India [16]. As per groundwater resource assessment carried out jointly by Central Ground Water Board and State Ground Water Organizations, as on 2009, annual groundwater withdrawal for irrigation has been estimated as 221 bcm, compared with 22 bcm for domestic and industrial [5].

Statewise details of groundwater extraction for irrigation and domestic and industrial uses [5] are given in Table 1.1, which clearly shows that Uttar Pradesh, Punjab, Madhya Pradesh, Maharashtra, Tamil Nadu, Rajasthan, Andhra Pradesh, Gujarat, and Haryana are the major groundwater withdrawal states for irrigation purpose. In the western part of the country, Gujarat and Maharashtra have seen a considerable spread of drip irrigation in some sizable pockets. In the south, Tamil Nadu and Andhra Pradesh are witnessing a dramatic increase in the area under drip irrigation and other precision farming methods and therefore these are the important states in the western and the southern parts of India.

**TABLE 1.1** Water Withdrawals for Various Purposes in Different States of India (bcm/year). (Source: http://pib.nic.in/newsite/erelease.aspx?relid=83055) )

| Sr. No. | States/union territories | Annual groundwater withdrawal (bcm/year) | | |
|---------|--------------------------|------------|-----------------------------|-------|
| | | Irrigation | Domestic and industrial uses | Total |
| 1 | Andhra Pradesh | 12.61 | 1.54 | 14.15 |
| 2 | Arunachal Pradesh | 0.002 | 0.001 | 0.003 |
| 3 | Assam | 5.333 | 0.69 | 6.026 |
| 4 | Bihar | 9.79 | 1.56 | 11.36 |
| 5 | Chhattisgarh | 3.08 | 0.52 | 3.60 |
| 6 | Delhi | 0.14 | 0.26 | 0.40 |
| 7 | Goa | 0.014 | 0.030 | 0.044 |
| 8 | Gujarat | 11.93 | 1.05 | 12.99 |
| 9 | Haryana | 11.71 | 0.72 | 12.43 |
| 10 | Himachal Pradesh | 0.23 | 0.08 | 0.31 |
| 11 | Jammu and Kashmir | 0.15 | 0.58 | 0.73 |
| 12 | Jharkhand | 1.17 | 0.44 | 1.61 |
| 13 | Karnataka | 9.01 | 1.00 | 10.01 |
| 14 | Kerala | 1.30 | 1.50 | 2.81 |
| 15 | Madhya Pradesh | 16.66 | 1.33 | 17.99 |
| 16 | Maharashtra | 15.91 | 1.04 | 16.95 |
| 17 | Manipur | 0.0033 | 0.0007 | 0.0040 |
| 18 | Meghalaya | 0.0015 | 0.0002 | 0.0017 |
| 19 | Mizoram | 0.000 | 0.0004 | 0.0004 |
| 20 | Nagaland | – | 0.008 | 0.008 |
| 21 | Orissa | 3.47 | 0.89 | 4.36 |
| 22 | Punjab | 33.97 | 0.69 | 34.66 |
| 23 | Rajasthan | 12.86 | 1.65 | 14.52 |
| 24 | Sikkim | 0.003 | 0.007 | 0.010 |
| 25 | Tamil Nadu | 14.71 | 1.85 | 16.56 |
| 26 | Tripura | 0.09 | 0.07 | 0.16 |
| 27 | Uttar Pradesh | 46.00 | 3.49 | 49.48 |
| 28 | Uttarakhand | 1.01 | 0.03 | 1.05 |
| 29 | West Bengal | 10.11 | 0.79 | 10.91 |
| **Total for states** | | **221.29** | **21.83** | **243.14** |
| **Total for union territories** | | **0.13** | **0.05** | **0.18** |
| **Grand total, India** | | **221.42** | **21.89** | **243.32** |

As agriculture utilizes more than 80% of the total water resources in India, it is imperative that efforts are focused on implementing water-saving technologies especially in the field of agriculture. Excessive draft has created problems especially in the overexploited regions of Punjab, Haryana, Rajasthan, and Gujarat. These problems include falling water tables, waterlogging and salinity, and inadequate access to safe drinking water and sanitation [6]. Irrigation efficiency is a ratio of volume of water required for consumptive use by the crop for its growth to the volume of water delivered from the source [1]. A basinwise study done at the Madras Institute of Development Studies estimated the overall irrigation efficiency in India to be 38%, which is quite low as compared to global standards. This is mainly because the efficiency of conventional flood irrigation technique, practiced in large parts of India, is low (35–40%) due to substantial conveyance and distribution losses [8]. Thus, MI techniques, including drip and sprinkler irrigation, have been introduced as water-conserving technologies in the past few decades in India. In drip irrigation, water is directly applied to the root zone of the crop in small quantities using a low-pressure delivery system with a network of pipes with small emitters (or drippers). This method helps to retain the soil moisture at consistent levels as against the flood irrigation method where there is a huge variation in soil moisture levels. Various field experiments have shown this technique to increase water use efficiency up to 80–90% depending on the crop and soil type [2, 11].

In spite of its advantages, the spread of MI has been restricted to only a few pockets across India. The main factors responsible for the limited spread of the technology have also been documented by quite a few researchers. These factors include:

1. High initial costs make the technology unfeasible especially for small and marginal farmers.
2. High emitter clogging rates due to dust and salinity.
3. Unsuitable cropping patterns. Drip Irrigation has been used for irrigating only a few selected crops in India. It is adopted mostly for coconut (19%), banana (11%), grapes (10%), mango (9.4%), citrus fruits (7.9%), and pomegranate (6.2%) [1].
4. It requires a lot of technical and management skills for setting up and upkeep.
5. Mechanical damage by farm labor, birds, and animals.
6. Easy availability of irrigation water, especially in northern parts of the country.

These factors have hindered the widespread use of this technology all over the country. Besides these, lack of demonstrability of the advantages at the field level may also be one of the reasons for the slow spread. Savings in energy may be particularly difficult to demonstrate. Moreover, some studies have also shown the costs of cultivation to increase due to high cost of management, use of improved quality of seeds, and increased fertilizer use to sustain increased yields [5, 7]. High rates of subsidies provided by the state and central governments (50–70% in most cases) have ensured that the technology is available to small farmers to some extent.

### 1.2.2   DEVELOPMENT AND SPREAD OF MICRO IRRIGATION

Experiments with MI technology were first conducted in Germany in the 1860s where water was pumped through clay pipes for irrigation. Current MI technology relates to the work of Simcha Blass of Israel in the 1930s. He accidentally discovered the concept and developed the first patented drip irrigation system. From Israel, the drip irrigation concept spread to Australia, North America, and South Africa by the late 1960s and eventually throughout the world.

The large-scale use of drip irrigation system started in 1970s in Australia, Israel, Mexico, New Zealand, South Africa, and the USA to irrigate vegetables and orchards and its coverage was 56,000 ha then [4]. The area under drip irrigation grew slowly but steadily and it was 0.41 million ha in 1981, 1.1 million ha in 1986, 1.77 million ha in 1991, 3.0 million ha in 2000, 6.2 million ha in 2006, and about 8.0 million ha in 2009 [4]. In India, Dr. R. K. Sivanappan started experimental studies in 1970 at Tamil Nadu Agricultural University in Coimbatore. The area under drip irrigation has increased from 1500 ha in 1985 to 70,859 ha in 1991–1992 and to 0.5 million ha in 2003 [2]. Table 1.2 shows spread of drip irrigation and its coverage with respect to the total area equipped for irrigation across various countries.

In India, drip irrigation is practiced using different kinds of systems such as conventional drip systems, indigenous pot and bucket drip systems, subsurface drip irrigation, family drip kits, and locally manufactured and assembled kits like Pepsee [14, 15]. The growth of MI in India over the years is shown in Figure 1.2. India, with a total arable area of 140 million ha with almost 50% of arable land irrigated, too has a huge

TABLE 1.2 Countrywise Coverage of Drip and Sprinkler Irrigation [2, 4].

| Sr. No. | Country | Total area equipped for irrigation (m-ha) | Sprinkler irrigation (ha) | Micro irrigation (ha) | Total sprinkler and micro irrigation | % of total irrigated area | Year of reporting |
|---|---|---|---|---|---|---|---|
| 1 | USA | 24.7 | 1,23,48,178.14 | 16,39,676.11 | 1,39,87,854.25 | 56.6% | 2009 |
| 2 | India | 60.9 | 30,44,940.00 | 18,97,280.00 | 49,42,220.00 | 8.1% | 2010 |
| 3 | China | 59.3 | 29,26,710.00 | 16,69,270.00 | 45,95,980.00 | 7.8% | 2009 |
| 4 | Russia | 4.5 | 35,00,000.00 | 20,000.00 | 35,20,000.00 | 78.2% | 2008 |
| 5 | Brazil | 4.45 | 24,13,008.00 | 3,27,866.00 | 27,40,874.00 | 61.6% | 2006 |
| 6 | Spain | 3.41 | 7,32,925.00 | 16,28,705.00 | 23,61,630.00 | 69.3% | 2010 |
| 7 | Italy | 2.67 | 9,81,163.00 | 5,70,568.00 | 15,51,731.00 | 58.1% | 2010 |
| 8 | France | 2.9 | 13,79,800.00 | 1,03,300.00 | 14,83,100.00 | 51.1% | 2011 |
| 9 | South Africa | 1.67 | 9,20,059.00 | 3,65,342.00 | 12,85,401.00 | 77.0% | 2007 |
| 10 | Saudi Arabia | 1.62 | 7,16,000.00 | 1,98,000.00 | 9,14,000.00 | 56.4% | 2004 |
| | Total | 211.89 | 3,50,72,650.10 | 1,00,82,729.10 | 4,51,55,379.30 | 21.3% | |

potential for MI, which is still underutilized. However, actual estimations for potential area under different studies show conflicting results.
While the Task Force on Micro Irrigation (2004) estimated a potential of
27 million ha for drip irrigation based on the area under crops, the Indian
Committee on Irrigation and Drainage (INCID) estimates a potential of
10.5 million ha [2]. That puts the market potential to anything between
1312 billion INR to 3375 billion INR.

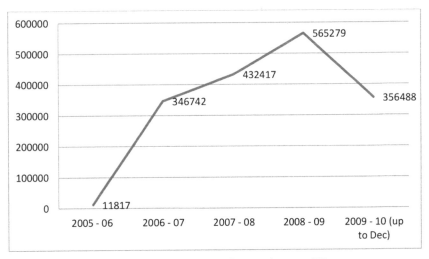

**FIGURE 1.2**    Growth of micro irrigation in India over the years [5].

The main factors responsible for the variation in spread may be: the
type of crops grown, the soil types in the region, availability of water
for irrigation, subsidies given by the state governments, and GOI. Often
ignored cause for the success or failure of drip irrigation is the marketing
and the effectiveness of the various marketing activities. MI is more suitable for widely spaced horticultural crops; plantation crops like bananas;
orchard crops like orange, grapes, pomegranate, flowers, vegetables; and
some other crops such as cotton, sugarcane, etc., which are grown on large
areas in Andhra Pradesh, Maharashtra, and Gujarat.

Therefore, there is a need to study and develop more consumer insights
to help marketers and policy makers to perform their tasks more efficiently
and effectively in the market while raising adoption rates for drip irrigation among farmers. This chapter presents the basic consumer insights for
marketers and policy makers alike.

## 1.3  METHODOLOGY

### 1.3.1  SAMPLING STATES

It has been found that the biggest hindrance in purchase of drip irrigation is the perception about drip irrigation. Therefore, a small perceptional survey was carried out across geographical areas and different agroclimatic zones to give us consumer data. The consumer preference data were analyzed to determine similarities and differences between disaggregated datasets for the adopters and nonadopters; and the same has been highlighted in this chapter to arrive at consumer insights for marketers and policy makers.

It is expected that perception differences can be accounted to reason other than informed perceptions. This study focuses on perception differences pertaining to drip irrigation but not perception based on rumors or such negative words and as such a sample from pockets of high adoption was chosen as the sampling methodology. Gujarat, Maharashtra, Andhra Pradesh, and Tamil Nadu states were chosen from the western and southern parts of India where maximum adoption for drip irrigation has been reported. Since all these four states are high adoption zones, the differences in perceptions can be attributed to information and awareness rather than marketing activities alone.

### 1.3.2  CHOICE OF POCKETS WITHIN EACH OF THE FOUR SAMPLING STATES

Within each state, further sampling was divided into various pockets so as to get maximum variation in terms of different crops and social settings. Two pockets were sampled in each chosen state. It was attempted to get variation in either the climatic profile or the crop profile of the drip application in the choice of districts in a state and across the states as well.

### 1.3.3  THE SURVEY METHODOLOGY

It was decided to administer the survey instrument at the farm level to the main farmer. The questionnaire had various sections on the profiles of farmer and farm, drip irrigation and its impact and performance. The data to be collected were largely perceptional apart from profile data, which were largely factual.

Within each district, two pockets or villages were sampled and the pockets or villages were chosen to have maximum variation in the crops and social settings and if possible market accessibility and distance from the dealers were also considered. Therefore, the sampling was able to cover 16 such pockets (4 states × 2 districts each × 2 pockets each). Thus, a minimum of 16 villages were covered for the survey. Within each pocket or village, the sampling design included about 25 adopters and 10 nonadopters across various landholding and caste classes if possible in each pocket identified within a district. The survey was conducted as planned across a total of 4 states and 16 identified pockets and a total of 499 respondents were administered the survey instrument. Their responses were collected and analyzed. The spread of the sample across the various states and adopters–nonadopters is given in Table 1.3.

**TABLE 1.3**    Sample Spread Across Four States.

|  | Andhra Pradesh | Gujarat | Maharashtra | Tamil Nadu | Subtotal |
| --- | --- | --- | --- | --- | --- |
| **Adopters** | 121(75.6%) | 76(70.3%) | 91(79.8%) | 82(69.5%) | 370(73.08%) |
| **Nonadopters** | 39(25.4%) | 32(29.7%) | 23(20.2%) | 35(30.5%) | 129(26.92%) |
| **Total** | 160 | 108 | 114 | 117 | 499 |

## 1.3.4   PROFILE OF A FARMER

### 1.3.4.1   THE AGE PROFILE

The basic survey unit was the farm and for each farm under survey the perception of the main farmer was recorded as responses. The profiling of farms and farmers was done to understand the secular features that could have impacted or influenced the responses of the farmers. A summary of some of the main profiling characteristics of the farms and the farmers and their households is given in this chapter. Table 1.4 presents the age profile of the overall sample and by the states. In Gujarat and Tamil Nadu, the age distribution curve shows a strong right-hand tilt.

**TABLE 1.4**  The Age Profile of the Overall Sample and By the States.

| Age (years) | Andhra Pradesh | Gujarat | Maharashtra | Tamil Nadu | Overall |
|---|---|---|---|---|---|
| | | | % | | |
| <30 | 17.5 | 11.1 | 13.2 | 1.7 | 10.87 |
| 31–40 | 39.4 | 18.5 | 40.4 | 6.8 | 26.27 |
| 41–50 | 26.3 | 35.2 | 26.3 | 36.4 | 31.05 |
| 51–60 | 13.1 | 21.3 | 15.8 | 44.9 | 23.77 |
| 61–70 | 3.8 | 11.1 | 3.5 | 6.8 | 6.3 |
| 71–80 | 0 | 2.8 | .9 | 2.5 | 1.55 |
| >80 | 0 | 0 | 0 | 0.8 | 0.2 |
| Total | 100 | 100 | 100 | 100 | 100 |
| Mean age | 41.36 | 47.28 | 41.74 | 51.6 | 45.14 |

## 1.3.4.2   EDUCATION PROFILE

Table 1.5 presents the education profile of the overall sample and for each state. About two-thirds of the respondents in the overall sample have been educated till middle school. The number of illiterate farmers was maximum in Andhra Pradesh where almost one-fifth of the farmers are illiterate, whereas almost the same proportion are graduates or have higher degrees in Gujarat and Maharashtra. Gujarat appears to have the maximum number of gentleman farmers with a balance of age and education mean age of 47.28 and with 22.20% having certified and higher education degrees.

**TABLE 1.5**  Education Profile of Farmer-Respondents (Statewise and Combined).

| Education | Andhra Pradesh (%) | Gujarat (%) | Maharashtra (%) | Tamil Nadu (%) | Overall (%) |
|---|---|---|---|---|---|
| Illiterate | 19.4 | 4.6 | 1.8 | 1.7 | 6.87 |
| Diploma/Certificate degree | 9.4 | 3.7 | 3.5 | 4.3 | 5.22 |
| Graduation or higher degree | 11.9 | 18.5 | 16.7 | 12.0 | 14.77 |
| High school | 20.0 | 13.0 | 14.9 | 15.4 | 15.82 |
| Middle school | 25.0 | 45.4 | 56.1 | 39.3 | 41.45 |
| Primary school | 14.4 | 14.8 | 7.0 | 27.4 | 15.9 |
| Total | 100.0 | 100.0 | 100.0 | 100.0 | 100.0 |

## 1.3.4.3   FAMILY SIZE PROFILE

Each of the state subsamples of the survey shows a variation in the average size of the family. In Table 1.6, Maharashtra shows the largest families with an average family size of 8.11 members per family and the minimum is in Andhra Pradesh at 5.25 members per family which is closer to the national average of India. The ratio of average family members and average drip irrigated areas is highest in Gujarat with a higher average family size of 6.39.

**TABLE 1.6**   Family Size Profile of Respondent Households (Statewise and Combined).

| Family size | Andhra Pradesh | Gujarat | Maharashtra | Tamil Nadu | Overall |
|---|---|---|---|---|---|
| Mean | 5.25 | 6.39 | 8.11 | 5.34 | 6.17 |
| Std. deviation | 2.07 | 3.13 | 4.20 | 1.62 | 3.07 |
| Minimum size | 2 | 2 | 3 | 1 | 1 |
| Maximum size | 16 | 19 | 26 | 10 | 26 |

## 1.3.4.4   LEADERSHIP PROFILE

A common belief is that villagers, who are more progressive and beneficial technology peers, often occupy village leadership positions and positions of importance and vice versa. It is also alleged that they often corner the benefits of schemes like the MI subsidy. Thus, responses were collected with respect to respondent's experience of being in or having been in a village leadership position. Table 1.7 collates these responses.

**TABLE 1.7**   Village Leadership Profile of Respondents (Statewise and Combined).

| Village leadership | Andhra Pradesh (%) | Gujarat (%) | Maharashtra (%) | Tamil Nadu (%) | Overall (%) |
|---|---|---|---|---|---|
| No | 96.3 | 86.1 | 84.2 | 97.0 | 91.3 |
| Yes | 3.8 | 13.9 | 15.8 | 3.0 | 8.7 |
| Total | 100.0 | 100.0 | 100.0 | 100.0 | 100.0 |

Around 9% of the respondents in the overall sample and state samples have held village leadership positions indicating a significant amount of leadership experience amidst the respondents. However, this ratio was very low in Andhra Pradesh.

### 1.3.4.5   CASTE PROFILE

Table 1.8 shows responses regarding the caste profile of the sample. In the overall sample, more than two-quarter of the respondents was from other backward class (OBC) and other minority. In terms of caste group, 63.2% of the respondents in Maharashtra are from the general category, while 94.5% of the respondents in Tamil Nadu are from OBCs. The other minorities constituted half of the sample in Andhra Pradesh compared with about two-thirds of the sample in Gujarat. This might contradict the popular notion that only "higher caste farmers" can afford drip irrigation. The low participation of the scheduled tribes as respondents is also noticeable in the sample. The surprising fact is the low participation of the scheduled castes (SCs) in the survey despite special incentives provided by most governments in terms of higher subsidy rates for SC farmers.

**TABLE 1.8**   Castewise Breakup of Respondents (Statewise and Combined).

| Caste | Andhra Pradesh (%) | Gujarat (%) | Maharashtra (%) | Tamil Nadu (%) | Overall (%) |
|---|---|---|---|---|---|
| **Scheduled castes** | 10.0 | 0.9 | 2.6 | 2.7 | 4.05 |
| **Scheduled tribes** | 0.6 | 0 | 4.4 | 0 | 1.25 |
| **Other backward classes** | 27.5 | 8.3 | 23.7 | 94.5 | 38.5 |
| **Other minority** | 58.1 | 67.6 | 6.1 | 2.7 | 33.65 |
| **General category** | 3.8 | 23.1 | 63.2 | 0 | 22.55 |
| **Total** | 100.0 | 100.0 | 100.0 | 100.0 | 100 |

### 1.3.4.6   FARM PROFILE

The landholding details are presented in Table 1.9. The Mean owned area is 7.53 acres across the sample and the leased-out area is nil, whereas the mean leased-in area stands at 4.9 acres. This again shows that the technology is either accepted by gentleman farmers or adopters transform into gentleman farmers with no leasing out and leasing beyond the land owned to increase the landholding under operation. The mean cultivated area across the sample is 7.68 acres that is much below the addition of the mean owned and leased-in areas. This probably signifies that drip irrigation is popular in regions where the smaller farms are able to lease-in land so that the average operated landholding is just above the average owned landholding. This usually is the case of regions where agriculture is prosperous in general. The mean irrigated area is 5.07 acres which is about two-thirds of the mean cultivated area for the sample. The mean drip irrigated area is 6.34 acres which is a higher value than the mean irrigated area. This clearly signifies that the larger farmers in the sample have higher drip irrigation adoption rates and hence also larger drip irrigated areas as compared to the smaller farmers. However, the bias seems to be reducing when compared to many of the earlier studies about a decade ago or so. The smaller farmers have also been able to afford and adopt drip irrigation, indicating very high standard deviation values which are almost equal the mean values. Thus, some very small farmers have also been able to adopt drip irrigation thereby reducing the landholding division between those who adopt and those who cannot or do not adopt drip irrigation technology. The minimum drip irrigated area for a respondent in the sample is just half an acre and this implies a lot for the availability, accessibility, and affordability of the technology for the smallholders.

**TABLE 1.9**   Landholding Profile of Farmer-Respondents (Statewise and Combined).

| Landholding (acres) | No. of households | Mean | Std. deviation | Minimum | Maximum |
|---|---|---|---|---|---|
| Owned area | 497 | 7.53 | 6.53 | 0.50 | 55.00 |
| Leased-in area | 19 | 4.90 | 4.49 | 1.00 | 16.00 |
| Leased-out area | 0 | 0 | 0 | 0 | 0 |
| Total cultivated area | 500 | 7.68 | 6.55 | 0.75 | 55.00 |
| Irrigated area | 318 | 5.07 | 6.48 | 0.50 | 95.00 |
| Rain-fed area | 46 | 4.24 | 4.02 | 1.0 | 15.0 |
| Drip irrigated area | 367 | 6.34 | 6.24 | 0.50 | 55.00 |

### 1.3.4.7   AREA PROFILE FOR DRIP IRRIGATION

In Table 1.10, the highest mean drip irrigated area per farmer is 7.46 acres in Gujarat followed by Maharashtra with 6.49 acres. The minimum is 4.63 acres in Andhra Pradesh and Tamil Nadu has a modest mean at 4.83 acres. Here again across all the states, the standard deviation is very high and is almost equal to the mean values.

**TABLE 1.10**   Drip Irrigated Area of Respondents (Statewise and Combined).

| Drip irrigated area (acres) | Andhra Pradesh | Gujarat | Maharashtra | Tamil Nadu | Overall |
|---|---|---|---|---|---|
| Mean | 4.63 | 7.46 | 6.49 | 4.83 | 6.34 |
| Standard deviation | 4.32 | 9.70 | 8.67 | 5.25 | 6.24 |

Thus, in almost all the four states, there are small holders, who have been able to adopt drip irrigation on their farms. At the same time, the higher than mean values of standard deviation in Gujarat and Maharashtra imply that there are some very large farmers who have adopted drip irrigation on very large areas, and for every such farmer there is a very small farmer who has adopted drip irrigation on a very limited lot of land. This certifies the claim that the technology is adopted by only the large farmers.

## 1.3.4.8   SOURCE OF WATER

Table 1.11 indicates the source of irrigation on the sampled farms. The common sources of irrigation were tube well, open well, canal as reported by 78.5%, 45%, and 12.4% respondents, respectively. Majority of the farmers had more than one source for irrigation water.

**TABLE 1.11**   Sources of Water on Respondent's Farms (Statewise and Combined).

| Sources of water | Andhra Pradesh (%) | Gujarat (%) | Maharashtra (%) | Tamil Nadu (%) | Overall (%) |
|---|---|---|---|---|---|
| Canal | 0.6 | 0.0 | 1.8 | 50.9 | 12.4 |
| Check dams | 0.0 | 0.0 | 0.0 | 2.6 | 0.6 |
| Lift from canal | 3.1 | 0.9 | 10.5 | 0.0 | 3.6 |
| Lift from stream/river | 0.6 | 0.0 | 2.6 | 1.7 | 1.2 |
| Lift from tanks | 0.0 | 0.0 | 1.8 | 0.0 | 0.4 |
| Open well | 3.1 | 43.5 | 97.4 | 52.6 | 45.0 |
| Tanks | 0.6 | 0.9 | 4.4 | 0.0 | 1.4 |
| Tube well | 95.0 | 58.3 | 74.6 | 78.4 | 78.5 |
| Other | 0.0 | 0.0 | 1.8 | 0.0 | 0.4 |

In Andhra Pradesh, Gujarat, and Tamil Nadu, the major source of irrigation water was tube well (95.0, 58.3, and 78.4%) followed by open well (3.1, 43.5, and 52.6%). In Maharashtra 97.4% of the farmers reported open well as the major source of irrigation; 74.6% and 10.5% of the respondents had their irrigation water requirement from tube wells and lifts from canal. There is a high degree of variation in the water sources for the sampled farms with some farmers accessing other sources as well.

## 1.4   RESULTS OF SURVEY

Thus, we find that the profiles of the respondents are very varied and no conclusions can be drawn with respect to the profile or adoption of drip irrigation based on the profile of the farms. This might go a long way in breaking many myths about drip irrigation adoption by only certain types of farmers: wealthy, large landholding, resource endowment, or "gentle-manly-ness." The sample shows only high variation profile and this indicates more equitable adoption than is commonly believed along any of the parameters indicated above.

### 1.4.1   DIFFERENCES OF OPINIONS OF ADOPTERS AND NONADOPTERS

This section reports the collated responses that highlight the differences in opinions of the adopters and nonadopters of drip irrigation across the four states and in specific states as well. These opinions give insights into the common myths and changes in these myths from the earlier studies and also how it is important to tone down the expectations of nonadopters as well in order to raise the satisfaction levels of future adopters of drip irrigation. While the insights can be used for marketing and promotions, they also ought to be used to decide on support and awareness activities to aid marketing efforts of drip irrigation systems. Table 1.12 shows assessment evaluation of drip irrigation systems by adopters and nonadopters; and includes assessment criteria (A to Q) to evaluate opinions of adopters and nonadopters of drip irrigation systems.

**TABLE 1.12** Assessment Evaluation of Drip Irrigation Systems by Adopters and Nonadopters.

| Evaluation parameter or opinion | Andhra Pradesh (%) | | Gujarat (%) | | Maharashtra (%) | | Tamil Nadu (%) | | Overall (%) | |
|---|---|---|---|---|---|---|---|---|---|---|
| | Adopter | Nonadopter | Adopter | Nonadopter | Adopter | Nonadopter | Adopter | Nonadopter | Adopter | Nonadopter |
| **A. Timely availability of water disaggregated by adopters and nonadopters** | | | | | | | | | | |
| Highly positive | 21.8 | 15.4 | 19.7 | 15.6 | 1.1 | 0.0 | 72.0 | 100 | 27.5 | 36.2 |
| Positive | 50.4 | 53.8 | 63.2 | 56.3 | 94.4 | 95.7 | 26.8 | 0.0 | 58.6 | 46.9 |
| No impact | 26.9 | 30.8 | 15.8 | 28.1 | 4.4 | 0.0 | 1.2 | 0.0 | 13.4 | 16.2 |
| Negative | 0,8 | 0.0 | 1.3 | 0.0 | 0.0 | 4.3 | 0.0 | 0.0 | 0.5 | 0.8 |
| Highly negative | 0.0 | 0.0 | 0.0 | 0.0 | 0.0 | 0.0 | 0.0 | 0.0 | 0.0 | 0.0 |
| **B. Increase in water table disaggregated by adopters and nonadopters** | | | | | | | | | | |
| Very high | **16.0** | **10.3** | **17.1** | **0.0** | 2.2 | 0.0 | 5.1 | 0.0 | **10.4** | **3.1** |
| High | 47.9 | 56.4 | 34.2 | 43.8 | 43.3 | 43.5 | **59.5** | **19.4** | 46.4 | 40.8 |
| Medium | 34.5 | 30.8 | 44.7 | 56.3 | 54.4 | 52.2 | **34.2** | **80.6** | 41.5 | 54.6 |
| Low | 1.7 | 2.6 | 3.9 | 0.0 | 0.0 | 4.3 | 0.0 | 0.0 | 1.4 | 1.5 |
| Very low | 0.0 | 0.0 | 0.0 | 0.0 | 0.0 | 0.0 | 1.3 | 0.0 | 0.3 | 0.0 |
| **C. Reduction in use of water quantity disaggregated by adopters and nonadopters** | | | | | | | | | | |
| Highly positive | 6.6 | 10.3 | **28.9** | **12.5** | 5.6 | 4.3 | **12.5** | **0.0** | **12.3** | **6.9** |
| Positive | 38.0 | 35.9 | 43.4 | 37.5 | 35.6 | 17.4 | 27.5 | 25.0 | 36.2 | 30.0 |
| No impact | 44.6 | 35.9 | 27.6 | 40.6 | 52.2 | 60.9 | 31.3 | 75.0 | 40.1 | 52.3 |
| Negative | **10.7** | **17.9** | **0.0** | **9.4** | **6.7** | **17.4** | **27.5** | **0.0** | 11.2 | 10.8 |
| Highly negative | 0.0 | 0.0 | 0.0 | 0.0 | 0.0 | 0.0 | 1.3 | 0.0 | 0.3 | 0.0 |

Assessment criteria

**TABLE 1.12** (Continued)

| Evaluation parameter or opinion | Andhra Pradesh (%) | | Gujarat (%) | | Maharashtra (%) | | Tamil Nadu (%) | | Overall (%) | |
|---|---|---|---|---|---|---|---|---|---|---|
| | Adopter | Nonadopter | Adopter | Nonadopter | Adopter | Nonadopter | Adopter | Nonadopter | Adopter | Nonadopter |
| **D. Misuse/abuse of water disaggregated by adopters and nonadopters** | | | | | | | | | | |
| Highly positive | 7.6 | 7.7 | 9.2 | 15.6 | 2.2 | 0.0 | 0.0 | 0.0 | 4.7 | 4.6 |
| Positive | 52.9 | 38.5 | 22.4 | 43.8 | 47.8 | 47.8 | 17.9 | 0.0 | 38.6 | 26.2 |
| No impact | 36.1 | 51.3 | 50.0 | 37.5 | 50.0 | 47.8 | 78.2 | 100 | 50.4 | 65.4 |
| Negative | 2.5 | 2.6 | 18.4 | 3.1 | 0.0 | 4.3 | 2.6 | 0.0 | 5.8 | 3.8 |
| Highly negative | 0.8 | 0.0 | 0.0 | 0.0 | 0.0 | 0.0 | 1.3 | 0.0 | 0.6 | 0.0 |
| **E. Impact on soil quality disaggregated by adopters and nonadopters** | | | | | | | | | | |
| Very high | 8.3 | 2.6 | 19.7 | 12.5 | 2.2 | 0.0 | 6.3 | 0.0 | 8.7 | 3.8 |
| High | 47.1 | 43.6 | 38.2 | 37.5 | 41.1 | 43.5 | 13.8 | 0.0 | 36.5 | 30.0 |
| Medium | 39.7 | 43.6 | 38.2 | 43.8 | 51.1 | 56.5 | 26.3 | 0.0 | 39.2 | 33.8 |
| Low | 4.1 | 10.3 | 0.0 | 0.0 | 5.6 | 0.0 | 35.0 | 0.0 | 10.4 | 3.1 |
| Very low | 0.8 | 0.0 | 3.9 | 6.3 | 0.0 | 0.0 | 18.8 | 100 | 5.2 | 29.2 |
| **F. Drip irrigation is suitable to all terrains** | | | | | | | | | | |
| Very high | 13.2 | 5.1 | 40.8 | 34.4 | 5.6 | 4.3 | 26.9 | 100 | 20.0 | 38.5 |
| High | 45.5 | 46.2 | 38.2 | 15.6 | 32.2 | 34.8 | 60.3 | 0.0 | 43.8 | 23.8 |
| Medium | 37.2 | 33.3 | 15.8 | 40.6 | 54.4 | 47.8 | 10.3 | 0.0 | 31.2 | 28.5 |
| Low | 4.1 | 15.4 | 3.9 | 3.1 | 7.8 | 13.0 | 1.3 | 0.0 | 4.4 | 7.7 |
| Very low | 0.0 | 0.0 | 1.3 | 6.3 | 0.0 | 0.0 | 1.3 | 0.0 | 0.5 | 1.5 |

TABLE 1.12  *(Continued)*

| Evaluation parameter or opinion | Andhra Pradesh (%) | | Gujarat (%) | | Maharashtra (%) | | Tamil Nadu (%) | | Overall (%) | |
|---|---|---|---|---|---|---|---|---|---|---|
| | Adopter | Nonadopter | Adopter | Nonadopter | Adopter | Nonadopter | Adopter | Nonadopter | Adopter | Nonadopter |
| **G. DI has met the varied irrigation needs successfully** | | | | | | | | | | |
| Highly satisfied | 5.0 | 2.6 | 19.7 | 31.3 | 1.1 | 0.0 | 53.8 | 100 | 17.8 | 36.2 |
| Satisfied | 58.8 | 56.4 | 47.7 | 28.1 | 11.1 | 21.7 | 30.0 | 0.0 | 38.4 | 27.7 |
| Undecided | 31.9 | 41.0 | 31.6 | 37.5 | 38.9 | 43.5 | 13.8 | 0.0 | 29.6 | 29.2 |
| Dissatisfied | 4.2 | 0.0 | 1.3 | 0.0 | 47.8 | 26.1 | 2.5 | 0.0 | 14.0 | 4.6 |
| Very dissatisfied | 0.0 | 0.0 | 0.0 | 3.1 | 1.1 | 8.7 | 0.0 | 0.0 | 0.3 | 2.3 |
| **H. Reduction in power requirement for irrigating the fields** | | | | | | | | | | |
| Very high | 5.8 | 7.7 | 25.0 | 12.5 | 4.4 | 0.0 | 13.6 | 0.0 | 11.1 | 5.3 |
| High | 49.6 | 46.2 | 44.7 | 25.0 | 27.8 | 17.4 | 34.6 | 0.0 | 39.9 | 23.1 |
| Medium | 34.7 | 41.0 | 26.3 | 46.9 | 57.8 | 78.3 | 24.7 | 0.0 | 36.4 | 37.7 |
| Low | 9.1 | 5.1 | 2.6 | 12.5 | 10.0 | 4.3 | 24.7 | 0.0 | 11.4 | 5.4 |
| Very low | 0.8 | 0.0 | 1.3 | 3.1 | 0.0 | 0.0 | 2.5 | 100 | 1.1 | 28.5 |
| **I. DI has helped in increasing the income of farmers** | | | | | | | | | | |
| Highly satisfied | 8.4 | 10.3 | 27.6 | 25.0 | 3.3 | 0.0 | 25.0 | 100 | 14.8 | 36.9 |
| Satisfied | 59.7 | 61.5 | 44.7 | 50.0 | 17.8 | 26.1 | 38.8 | 0.0 | 41.6 | 35.4 |
| Undecided | 27.7 | 25.6 | 23.7 | 15.6 | 53.3 | 52.2 | 35.0 | 0.0 | 34.8 | 20.8 |
| Dissatisfied | 4.2 | 2.6 | 3.9 | 6.3 | 21.1 | 21.7 | 1.3 | 0.0 | 7.7 | 6.2 |
| Very dissatisfied | 0.0 | 0.0 | 0.0 | 3.1 | 4.4 | 0.0 | 0.0 | 0.0 | 1.1 | 0.8 |

**TABLE 1.12** (Continued)

| Evaluation parameter or opinion | Andhra Pradesh (%) | | Gujarat (%) | | Maharashtra (%) | | Tamil Nadu (%) | | Overall (%) | |
|---|---|---|---|---|---|---|---|---|---|---|
| | Adopter | Nonadopter | Adopter | Nonadopter | Adopter | Nonadopter | Adopter | Nonadopter | Adopter | Nonadopter |
| **J. Better market power when dealing with traders** | | | | | | | | | | |
| Highly positive | 6.7 | 10.3 | 14.5 | 6.3 | 3.3 | 0.0 | 2.6 | 0.0 | 6.6 | 4.6 |
| Positive | 51.3 | 35.9 | 34.2 | 25.0 | 48.9 | 43.5 | 10.3 | 0.0 | 38.3 | 24.6 |
| No impact | 37.0 | 43.6 | 48.7 | 68.8 | 45.6 | 56.5 | 73.1 | 100 | 49.3 | 67.7 |
| Negative | 5.0 | 10.3 | 2.6 | 0.0 | 2.2 | 0.0 | 14.1 | 0.0 | 5.8 | 3.1 |
| Highly negative | 0.0 | 0.0 | 0.0 | 0.0 | 0.0 | 0.0 | 0.0 | 0.0 | 0.0 | 0.0 |
| **K. Better social status** | | | | | | | | | | |
| Highly positive | 2.5 | 2.6 | 27.6 | 12.5 | 3.3 | 0.0 | 7.3 | 0.0 | 9.0 | 3.8 |
| Positive | 53.8 | 64.1 | 50.0 | 28.1 | 46.7 | 30.4 | 47.6 | 100 | 49.9 | 59.2 |
| No impact | 41.2 | 25.6 | 22.4 | 56.3 | 48.9 | 69.6 | 32.9 | 0.0 | 37.3 | 33.8 |
| Negative | 2.5 | 7.7 | 0.0 | 3.1 | 1.1 | 0.0 | 11.0 | 0.0 | 3.5 | 3.1 |
| Highly negative | 0.0 | 0.0 | 0.0 | 0.0 | 0.0 | 0.0 | 1.2 | 0.0 | 0.3 | 0.0 |
| **L. After sales, service is costly** | | | | | | | | | | |
| Very high | 9.9 | 7.7 | 19.7 | 9.4 | 4.4 | 8.7 | 5.1 | 0.0 | 9.6 | 6.2 |
| High | 35.5 | 25.6 | 18.4 | 34.4 | 24.4 | 21.7 | 35.4 | 0.0 | 29.2 | 20 |
| Medium | 43.0 | 51.3 | 26.3 | 40.6 | 56.7 | 56.5 | 40.5 | 100 | 42.3 | 63.1 |
| Low | 9.1 | 15.4 | 18.4 | 12.5 | 13.3 | 13.0 | 15.2 | 0.0 | 13.4 | 10.0 |
| Very low | 2.5 | 0.0 | 16.9 | 3.1 | 1.1 | 0.0 | 3.8 | 0.0 | 5.5 | 0.8 |

**TABLE 1.12** (*Continued*)

| Evaluation parameter or opinion | Andhra Pradesh (%) Adopter | Andhra Pradesh (%) Nonadopter | Gujarat (%) Adopter | Gujarat (%) Nonadopter | Maharashtra (%) Adopter | Maharashtra (%) Nonadopter | Tamil Nadu (%) Adopter | Tamil Nadu (%) Nonadopter | Overall (%) Adopter | Overall (%) Nonadopter |
|---|---|---|---|---|---|---|---|---|---|---|
| **M. Good quality service, after sales** | | | | | | | | | | |
| Very high | 6.7 | 2.6 | 23.7 | 9.4 | 0.0 | 0.0 | 9.8 | 0.0 | 9.2 | 3.1 |
| High | 47.5 | 28.2 | 32.9 | 34.4 | 30.0 | 26.1 | 8.5 | 0.0 | 31.5 | 21.5 |
| Medium | 35.8 | 53.8 | 21.1 | 53.1 | 58.9 | 56.5 | 26.8 | 0.0 | 36.4 | 39.2 |
| Low | 9.2 | 15.4 | 14.5 | 0.0 | 7.8 | 13.0 | 43.9 | 36.1 | 17.7 | 16.9 |
| Very low | 0.8 | 0.0 | 7.9 | 3.1 | 3.3 | 4.3 | 11.0 | 63.9 | 5.2 | 19.2 |
| **N. Clogging of DI is a big problem** | | | | | | | | | | |
| Very high | 15.7 | 15.4 | 19.7 | 15.6 | 4.4 | 4.3 | 13.6 | 0.0 | 13.3 | 9.2 |
| High | 38 | 35.9 | 14.5 | 21.9 | 33.3 | 26.1 | 19.8 | 0.0 | 28.0 | 20.8 |
| Medium | 38.8 | 38.5 | 35.5 | 53.1 | 53.3 | 65.2 | 12.3 | 19.4 | 35.9 | 41.5 |
| Low | 5.8 | 10.3 | 21.1 | 9.4 | 8.9 | 4.3 | 53.1 | 0.0 | 20.1 | 6.2 |
| Very low | 1.7 | 0.0 | 9.2 | 0.0 | 0.0 | 0.0 | 1.2 | 80.6 | 2.7 | 22.3 |
| **O. DI equipment is easy to service and maintain** | | | | | | | | | | |
| Very high | 15.7 | 15.4 | 25.0 | 9.4 | 1.1 | 0.0 | 42.7 | 58.3 | 20.1 | 23.1 |
| High | 40.5 | 30.8 | 31.6 | 40.6 | 20.0 | 34.8 | 48.8 | 41.7 | 35.5 | 36.9 |
| Medium | 36.4 | 46.2 | 30.3 | 46.9 | 72.2 | 56.5 | 6.1 | 0.0 | 37.1 | 35.4 |
| Low | 5.0 | 7.7 | 10.5 | 3.1 | 5.6 | 8.7 | 1.2 | 0.0 | 5.4 | 4.6 |
| Very low | 2.5 | 0.0 | 2.6 | 0.0 | 1.1 | 0.0 | 1.2 | 0.0 | 1.9 | 0.0 |

**TABLE 1.12** (Continued)

| Evaluation parameter or opinion | Andhra Pradesh (%) | | Gujarat (%) | | Maharashtra (%) | | Tamil Nadu (%) | | Overall (%) | |
|---|---|---|---|---|---|---|---|---|---|---|
| | Adopter | Nonadopter | Adopter | Nonadopter | Adopter | Nonadopter | Adopter | Nonadopter | Adopter | Nonadopter |
| **P. DI is difficult to master** | | | | | | | | | | |
| Very high | 5.0 | 2.6 | 7.9 | 15.6 | 0.0 | 0.0 | 5.1 | 25.0 | 4.4 | 11.5 |
| High | 28.1 | 28.2 | 13.2 | 25.0 | 8.9 | 13.0 | 5.1 | 0.0 | 15.3 | 16.9 |
| Medium | 51.2 | 48.7 | 25.0 | 40.6 | 34.4 | 39.1 | 12.8 | 11.1 | 33.4 | 34.6 |
| Low | 12.4 | 15.4 | 28.9 | 3.1 | 33.3 | 26.1 | 44.9 | 63.9 | 27.9 | 27.7 |
| Very low | 3.3 | 5.1 | 25.0 | 15.6 | 23.3 | 21.7 | 32.1 | 0.0 | 18.9 | 9.2 |
| **Q. DI is very costly** | | | | | | | | | | |
| Very high | 9.1 | 10.3 | 23.7 | 21.9 | 4.4 | 13.0 | 12.2 | 0.0 | 11.7 | 10.8 |
| High | 34.7 | 38.5 | 30.3 | 28.1 | 12.2 | 8.7 | 11.0 | 0.0 | 23.0 | 20.0 |
| Medium | 33.9 | 35.9 | 38.2 | 50.0 | 63.3 | 78.3 | 50.0 | 100 | 45.5 | 64.6 |
| Low | 19.0 | 10.3 | 3.9 | 0.0 | 14.4 | 0.0 | 20.7 | 0.0 | 15.2 | 3.1 |
| Very low | 3.3 | 5.1 | 3.9 | 0.0 | 5.6 | 0.0 | 6.1 | 0.0 | 4.6 | 1.5 |

### 1.4.2.1  TIMELY AVAILABILITY OF WATER

Table 1.12A shows the responses to the impact on timely water availability due to drip irrigation. Overall, the response is more positive for the nonadopters than for the adopters, due to overwhelming expectation among nonadopters in Tamil Nadu. The responses have no perceivable difference in Maharashtra and the adopters are slightly more positive about the impacts than the nonadopters in states of Andhra Pradesh and Gujarat. This benefit of drip irrigation can be used for marketing in these two states, whereas the expectations will need to be toned down in Tamil Nadu.

### 1.4.2.2  IMPACT OF DRIP IRRIGATION ON WATER TABLE INCREASE

Table 1.12B shows that there are significant differences in opinions with regard to the impact of drip irrigation on water table increase. Overall, the nonadopters rate the impact on water table increase slightly lower than that by adopters. This trend is seen mildly in Andhra Pradesh and more strongly in Gujarat. However, there is no perceivable difference in the perceptions in Maharashtra and the nonadopters rate the impact much higher in Tamil Nadu than by the adopters. Thus, a diametrically opposite communication is needed for marketing drip irrigation in Andhra Pradesh, Gujarat, and Tamil Nadu, whereas this is not a differentiating aspect for the state of Maharashtra.

### 1.4.2.3  IMPACT OF DRIP IRRIGATION ON REDUCTION IN WATER QUANTITY

Table 1.12C presents the collated responses to the perception of impact of drip irrigation on reduction in water quantity used for irrigation. The results show that the adopters rate drip irrigation very favorably in terms of reduction in use of water quantity compared with nonadopters. At the same time, there are many nonadopters who rate it very poor on reduction of water quantity used. This signals that marketing drip irrigation as a water-saving technology is not enough but it ought to be marketed as a technology that allows the farmer to do better agriculture even under limited quantities of water. However, there is a segment of adopters in Tamil Nadu who do not rate drip irrigation favorably in terms of reduction in water use. Therefore, in Tamil Nadu the communication has to be more even and balanced and the specific details of this segment needs to be identified.

### 1.4.2.4   IMPACT OF DRIP IRRIGATION ON THE MISUSE/ABUSE OF WATER

Table 1.12D shows the perception of impact of drip irrigation on the misuse/abuse of water. The adopters are more favorable toward drip irrigation on this aspect, in the southern states of Andhra Pradesh and Tamil Nadu. However, there is no perceptible difference in Maharashtra; whereas in Gujarat, the nonadopters are more favorable for drip irrigation on this aspect compared with the adopters. This again splits the market and the need is to prepare different communication and marketing strategy in different states. This aspect does not appear to be a useful aspect for marketing of drip irrigation for Maharashtra.

### 1.4.2.5   IMPACT ON SOIL QUALITY

Table 1.12E shows the perception about the impact on soil quality. In the states of Andhra Pradesh and Gujarat, the adopters have rated drip irrigation slightly more positive in terms of impact on soil quality. This difference is noticeable and stronger in Andhra Pradesh compared with Gujarat, whereas in Maharashtra there is no perceptible impact. On the other hand, in Tamil Nadu, each and every nonadopter has rated drip irrigation as very poor in terms of impact on soil quality. This clearly highlights one of the strongest beliefs that stop the nonadopters from opting for drip irrigation. This is a very strong opinion and must be taken into consideration by all drip irrigation marketers in Tamil Nadu.

### 1.4.2.6   SUITABILITY OF DRIP IRRIGATION TO ALL TERRAINS

Table 1.12F gives us the responses of adopters and nonadopter respondents about their perception of suitability of drip irrigation to all terrains.

Nonadopters have very strong positive perception on this aspect in Tamil Nadu. However, this seems a myth as it is so far from that reported by adopters. Nevertheless, it is an important perception that needs to be kept in mind by the marketers. The ratings for drip irrigation are much higher for the adopters than the nonadopters in other states especially Andhra Pradesh and Gujarat. The positive perception of drip irrigation is much higher.

### 1.4.2.7   WHETHER DRIP IRRIGATION MEETS THE VARIED NEEDS OF IRRIGATION SUCCESSFULLY OR NOT

Table 1.12G presents the responses to the question whether drip irrigation has met the varied needs of irrigation successfully or not. Once again, one can see that the perception in favor of drip irrigation is more strong for the nonadopters than the adopters especially in the states of Tamil Nadu, Gujarat and more mildly in Maharashtra. This clearly highlights as one of the myths about drip irrigation. However, the reality is much different. This can be problematic as a more positive rating by nonadopters could mean very high expectations reducing the satisfaction levels of future adopters with the technology. However, Andhra Pradesh bucks this trend and the nonadopters here appear to be more balanced in their perception of drip irrigation on this aspect.

### 1.4.2.8   REDUCTION IN POWER USED FOR IRRIGATION

Table 1.12H indicates the perceptions that drip irrigation leads to reduction in power use for irrigation. The ratings of adopters are consistently higher than those of the nonadopters across the states of Gujarat, Maharashtra, and Tamil Nadu and there is no perceptible difference in the ratings of respondents in Andhra Pradesh. This is clearly one of the aspects on which positive marketing can be done. However, the marketing has to be cleverly designed as pricing of power for agriculture is itself highly subsidized and therefore not a matter of great concern for the average farmer in the country. This may need a social marketing approach, which can boost sales of drip irrigation.

### 1.4.2.9   INCREASING THE INCOME OF FARMERS

Table 1.12I shows the impact of drip irrigation on increasing the income of the respondents. The nonadopters in Tamil Nadu feel very strongly that drip irrigation contributes to an increase in income while there is almost no perceptible difference in neighboring Andhra Pradesh among the adopters and nonadopters. However, the adopter respondents in Gujarat have rated drip irrigation more favorably than their nonadopter counterparts. In Maharashtra, most of the respondents show no difference; however, there are small segments of adopters which rate drip irrigation much higher and much lower than nonadopters in its impact on increasing their incomes. There is a clear need to dispel the myths about increasing income that drip irrigation by its mere application alone will not result in an increase in income but it can enable changes in agriculture that can easily be a better market power when dealing with traders to increase in incomes.

### 1.4.2.10   BETTER MARKET POWER

Table 1.12J shows that adopter-respondents in Andhra Pradesh, Gujarat, and Maharashtra (in respect of strength of perception) felt that drip irrigation leads to better market power when dealing with traders as compared with the perception of nonadopters. However, in Tamil Nadu, most of the respondents—adopters and nonadopters—alike felt that drip irrigation did not have any impact on the market power of farmers in dealing with the traders. However, a small but significant section of one-seventh of adopter-respondents felt that it had a positive impact. At the same time about one-sixth of the adopter-respondents felt that it had a negative impact on market power when dealing with traders.

### 1.4.2.11   ENHANCEMENT OF SOCIAL STATUS

In terms of impact on social status, Table 1.12K presents that the adopters in states of Gujarat and Maharashtra strongly felt that drip irrigation led to an enhancement of social status compared with nonadopters. However, the nonadopters rated drip irrigation more favorably in Andhra Pradesh and somewhat also in Tamil Nadu. Thus, there is a clear difference between the western and the southern states on this aspect. However, it signals that activities that allow this intangible benefit to be harnessed by adopters need to be undertaken in Andhra Pradesh and Tamil Nadu as the nonadopters are attracted toward drip for this intangible benefit.

### 1.4.2.12   COST OF AFTER-SALES SERVICE

Table 1.12L gives the responses on the cost of after-sales service. Apart from the state of Gujarat, the adopters found the after-sales service to be costly compared to the perceptional rating given by the nonadopters. This is an important aspect of sales of drip irrigation and marketers need to focus on bringing down the after-sales service costs as it seems to hamper the perception of the product post-adoption. The marketers need to focus on informing the prospective customers on ways to cut down after-sales costs and how good service is available to them.

### 1.4.2.13   QUALITY SERVICES AFTER SALES

Table 1.12M shows us that the adopters across states perceive after-sales service for drip irrigation to be of good quality. This confirms the opinion that there is a need to market the availability of good quality after-sales service for drip irrigation as a selling point by the marketers. It is also likely that the market will differentiate one seller from others on this aspect.

### 1.4.2.14   CLOGGING OF DRIP IRRIGATION

In Table 1.12N, one can see that in general adopters responded that clogging of drip irrigation systems is a big problem. This perception difference between the adopters and nonadopters is most balanced in Andhra Pradesh. There is a significant section of adopter-farmers in Gujarat that do not confirm to this trend. Thus, it is clear that marketers need to be prepared to handhold future customers on clogging issues as they do not perceive them as much at the time of purchase. At the same time, the marketers can prepare educational materials to make the prospective customers more aware about the issue and also better prepared at dealing with it. At the same time, another interpretation possible is that despite the difference it is not an important issue for purchase as the adopters are rating it as a big issue.

### 1.4.2.15   DI EQUIPMENT IS EASY TO SERVICE AND MAINTAIN

We find similar trends with the perception that drip irrigation equipment is easy to maintain as shown in Table 1.12O.

### 1.4.2.16   DRIP IRRIGATION IS DIFFICULT TO MASTER

Table 1.12P shows that drip irrigation is perceived as difficult to master by most nonadopters across states. In Andhra Pradesh, the perception is similar to that of the adopters. This is clearly an important aspect that the marketers need to communicate with prospective customers to boost sales.

### 1.4.2.17   DRIP IRRIGATION IS EXPENSIVE

Table 1.12Q collates the responses to the perception that drip irrigation is very costly. There is a clear perception difference as there is larger proportion of adopters who disagree with this compared with nonadopters. This is especially strong in Tamil Nadu and Maharashtra followed by Gujarat in order of proportions. At the same time, there is a small but significant segment in Maharashtra where adopters feel that drip irrigation is very costly as compared with the perception of nonadopters. This proportion is slightly higher in Tamil Nadu. Thus, a more segmented approach within each state needs to be taken on this aspect of cost. Therefore, multiple marketing activities using different payment modes should be used for selling drip irrigation system.

## 1.5   SUMMARY

Drip irrigation is a technology with a very high potential in India. Farmers in India have already proven that they can adopt and use this technology to their advantage across various states and agroclimatic zones. Thus, there is no doubt on the usefulness of this technology. However, its progress has been slow and marketers need more consumer insights to guide them to market drip irrigation better to boost its adoption. This chapter highlights some consumer perceptions and focuses on the differences between the adopters and nonadopters.

The results of survey indicate that there are many myths about drip irrigation and these usually make the nonadopters have very high expectations from the technology and thereby suppress the satisfaction levels post-adoption. Marketing needs to take into consideration such myths and educate the prospective customers about these and also take up activities to ensure that customers are better prepared and taken care of better to deal with the actual events post-adoption. Such myths include that drip irrigation is a panacea for all water availability issues in irrigation; adoption of drip irrigation will lead to higher income; drip irrigation is suitable to all terrains; drip irrigation can fulfil the various irrigation requirements successfully, etc. There are lots of myths about drip irrigation prevailing with the nonadopters. These appear to be more serious issues than lack of knowledge of various other benefits of drip irrigation. This is partially due to the sampling in high adoption pockets. These are still very important for drip irrigation marketers to take into consideration as the myths prevail even in areas of high adoption and hence they are certain to impact the post-purchase satisfaction and subsequent adoption decisions adversely.

There are many other aspects of drip irrigation that marketers need to focus: drip irrigation helps in timely and adequate availability of irrigation; it is costly and difficult to master; it enhances the chances for increasing incomes and increase in water tables, and also enables agriculture with very limited water availability while effecting a reduction in the usage of water and power consumed for irrigation and provision of very good quality after-sales service.

These two activities of myth busting and imparting awareness are not uniform across India. While some are found to be true across all the four sampled states, some pertain to only southern or western states showing a regional bias and others still are present in only one state and at times even at substate level (the same has not been reported here as that is not the focus of this chapter). Thus, customization in the marketing activities and communication has to be the primary focus for marketers. They cannot take the liberty of equating one market with another without paying the price for the same.

The two major activities of myth-busting and imparting awareness need to be backed up by operational activities that enable efficient land-holding of the farmer to be able to change his agriculture as enabled by drip irrigation to earn more, save water and power, and also get the best agricultural and market advice for the same backed by efficient after-sales service and education to master the usage of drip irrigation.

## KEYWORDS

- after-sales service
- drip irrigation
- micro irrigation
- adopters
- nonadopters

## REFERENCES

1. GOI; *Report of the Inter Ministry Task Group on Efficient Utilization of Water Resources*; Planning Commission—GOI, 2004.
2. INCID; *Drip Irrigation in India*; Indian National Committee on Irrigation and Drainage: New Delhi, 1994.
3. Keller, J.; Blisner, R. D. *Sprinkler and Trickle Irrigation*. Chapman and Hall: New York, 1990, pp. 325.
4. Kulkarni, S. A.; Reinders, F. B.; Ligetvari, F.. In *Global Scenario of Sprinkler and Micro Irrigated Areas*, 7th International Micro Irrigation Congress, Sep 10–16, 2006, PWTC, Kuala Lumpur, Malaysia.
5. NABARD, Growth of micro irrigation in India over the years. In: Evaluation Study of Centrally Sponsored Schemes on Micro Irrigation (Nabcons); Unpublished Report NCPAH (National Committee on Plasticulture Applications in Horticulture) – Government of India; Consultancy Services Private Limited, New Delhi, 2009.
6. Narain, V. India's Water Crisis: The Challenge of Governance. *Water Policy*. **2000,** 2(6), 433–444.
7. Narayanamoorthy, A. *Economics of Drip Irrigation: A Comparative Study of Maharashtra and Tamil Nadu*: AERC, Pune, 1997.
8. Narayanamoorthy, A. Averting Water Crisis by Drip Method of Irrigation: A Study of Two Water-Intensive Crops. *Indian J. Agr. Eco.*, **2003,** 58(3), 427–437.
9. Phansalkar, S. J. ; Verma, S., India's Water Future 2050: Potential Deviations from Business as Usual. *Int. J. Rur. Mngm*. IRMA, Anand – Gujrat, **2007.**

10. Punetha, A.; Reddy, K. Y. *APMIP: the First and Largest Comprehensive Micro Irrigation Project in India.* 7th International Micro Irrigation Congress, Sep 10–16, 2006, PWTC, Kuala Lumpur, Malaysia.

11. Sivanappan, R. K. Prospects of Micro Irrigation in India. *Irrig. Drain. Syst.*, **1994,** *8*, 49–58.

12. UNEP 2012. Accessed from: http://www.unep.org/training/programmes/Instructor% 20Version/Part_2/Activities/Economics_of_Ecosystems/Water/Supplemental/ Global_Water_Resources.pdf

13. UNESCO. *World Water Development Report: IV* (March). World Water Assessment Programme of UNESCO, 2012.

14. Varma, S.; Verma, S.; Namara R. E. *IWMI Water Policy Briefing*, **2006,** *23.*.

15. Verma, S.. *Promoting Micro Irrigation in India: A Review of Evidence and Recent Developments*; IWMI, 2004.

16. World Bank. *Deep Wells and Prudence: Towards Pragmatic Action for Addressing Groundwater Overexploitation in India*, 2010.

# CHAPTER 2

# POLICY OPTIONS FOR BETTER IMPLEMENTATION OF MICRO IRRIGATION: CASE STUDY OF INDIA

K. PALANISAMI[1*], K. KRISHNA REDDY[2], S. RAMAN[3], and T. MOHANASUNDARI[4]

[1]*International Water Management Institute (IWMI), ICRISAT Campus, 401/5, Patancheru 502324, Medak District, Telangana, India*

[2]*International Water Management Institute (IWMI), ICRISAT Campus, 401/5, Patancheru 502324, Medak District, Telangana, India*

[3]*Water Resources Expert and Consultant, Mumbai, India*

[4]*Tamil Nadu Agricultural University, Coimbatore, Tamil Nadu, India*

*Corresponding author. E-mail: k.palanisami@cgiar.org*

## CONTENTS

## ABSTRACT

The present study was undertaken in five selected states and indicated higher rate of returns for MI (ranging from 14% to 67%) among different farm categories across the states. The major constraints in expanding the MI area as reported by the farmers are: higher unit cost of system, lack of technical knowhow on fertigation and time lag, and higher transaction cost in getting the MI subsidy.

## 2.1 INTRODUCTION

The capacity and capabilities of India to efficiently develop and manage water resources are key determinants for global food security in the 21st century [16]. In India, almost all possible ways for viable irrigation potential have already been tapped. However, the water demand for different sectors has been growing continuously [15, 19] and the demand management becomes overall key strategy for managing scarce water resources [6]. Since agriculture is the major water consuming sector in India, the demand management in agriculture in water-scarce and water-stressed regions is central to reduce the aggregate demand for water to match the available future supplies [18]. Various options are available for reducing water demand in agriculture:

- Firstly, the supply side management practices include watershed development and water resource development through major, medium, and minor irrigation projects.
- The second is through the demand management practices, which include improved water management technologies/practices.

Concerned by growing water scarcity and the need to use the available water more efficiently, the Government of India (GOI) has been making efforts to improve water use efficiency (WUE) by encouraging use of micro irrigation (MI) (drip and sprinkler) technologies. The MI technologies such as drip and sprinkler, are key interventions in water saving and improving the crop productivity. Evidences show that the water can be saved up to 40–80% and WUE can be enhanced in a properly designed and managed MI system compared to 30–40% under conventional practice [4, 17 cited in 18]. The successful adoption of MI requires, in addition to technical and economic efficiency, two additional preconditions, viz., technical knowledge about the technologies and accessibility of technologies through institutional support systems [7].

This chapter outlines new approaches/models that can help to boost the MI expansion in India. The major objectives of this chapter are to examine the pathways for expanding the MI area in India:

- To examine the economics of MI adoption in different regions;
- To identify the major constraints in expanding the MI adoption; and
- To suggest suitable implementation models for upscaling MI in India.

## 2.2  HISTORY OF MICRO IRRIGATION SUBSIDY: INDIA

Keeping the goal of improving the WUE by expanding MI area in the country as well as farmers' profitability, subsidizing farmers' capital cost of MI systems is considered important. In 2004, the importance of promoting MI adoption largely started with the recommendations of the *Task Force on Micro Irrigation* (TFMI), which recommended more financial resources for subsidies, with state governments taking up to 10% of the cost, while the funds from GOI would account for 40% and advised greater flexibility for states to determine their appropriate implementation structure and institutional mechanisms for subsidy disbursement [13]. Based on the recommendations of TFMI in 2006, the Central Sponsored Scheme (CSS) on MI was launched. Based on the success of CSS and realizing the importance of MI in India, the scheme was upgraded to a Mission mode in 2010 under the name of *National Mission on Micro Irrigation* (NMMI) [9]. The operational guidelines for NMMI stresses that "the success of the scheme will depend on an effective delivery mechanism." In the NMMI, the subsidy was 50% and the state governments were required to contribute another 10%. The subsidy was uniform for all the types of beneficiaries.

Subsequently, the GOI felt the need for revising the subsidy norms for the MI as suggested earlier by NMMI. By 2014, the NMMI was subsumed into *National Mission on Sustainable Agriculture* (NMSA) with *On Farm Water Management* specifically focusing on MI aspects of the program [10, 11]. In the NMSA, the subsidy norms have been revised as follows [11]: The subsidy will be 35% of cost of installation for small and marginal farmers and 25% of cost of installation for others under the non-*Drought Prone Area Program* (DPAP)/*Desert Development Program* (DDP)/and North Eastern & Himalayan (NE&H) states. In the DPAP/DPP and NE&H states the corresponding subsidy rates are fixed at 50% and 35%, respectively [11]. The scheme has undergone further changes with the introduction of *Pradhan Manthri Krishi Sinchayee Yojana* (PMKSY), which focuses on "water to every farm" and shift of focus from "more crop per acre" to "more crop per drop." PMKSY envisages wholesome and inclusive developmental approach to the subject of water in agriculture. District-level and state-level irrigation plans will have to be prepared and wetted with the PMKSY at GOI and implement the plan. Funds available under various schemes will be dovetailed into PMKSY. It is a multidisciplinary integrated approach to water management [3]. The scheme was initiated in May 2015 and is yet to take full shape.

Therefore, it is important to see how best the MI area in the country can be upscaled given the slow adoption levels. It is anticipated that MI adoption can be increased substantially by either improving the performance of existing implementation mechanism and/or by introducing new approaches for providing incentives or subsidies to farmers.

## 2.3 METHODOLOGY: SAMPLE AND DATA

This chapter uses the data collected during 2010 and updated in 2014 for five states (viz., Andhra Pradesh, Gujarat, Karnataka, Maharashtra, and Tamil Nadu). Both secondary and primary data were collected. The secondary data were collected covering the state-level MI sources, cropping pattern, existing area under MI, and the government subsidy details. The primary data were collected from 150 farmers from each selected state using semi-structured questionnaire covering source of irrigation, farm size, irrigated area, area under MI, crops grown, subsidy availed, crop income, and expenditure under crops with and without MI. Farmlevel constraints for adoption of MI and suggestions for better adoption were also obtained from the field surveys. The sample was post stratified into marginal, small, and large farmers. Key variables, such as MI area and net income under different farm categories, were updated using the 2014 data collected exclusively for this purpose. The internal rate of return (IRR) due to MI was worked out using the annualized capital cost of the system, average life of the MI system, and the additional crop income that will occur during the life period of the MI system in the farm.

$$\text{Annualized cost of MI} = [(\text{capital cost of MI}) *$$
$$(1 + i)^{AL} * i] \div [(1 + i)^{AL} - 1], \tag{2.1}$$

where AL = average life of MI system (8 years); and $i$ = discount rate (10%).

## 2.4  RESULTS AND DISCUSSION

### 2.4.1  MICRO IRRIGATION ADOPTION BY VARIOUS FARM CATEGORIES

The estimated potential of MI in the country is about 42 million ha [14] and the current coverage of area under MI in the country (as on March 2013) is only about 6.6 million ha. Recently, it is in the process of revision according to which the potential is estimated to be 47 million hectares [8]. The percentage of actual area against the revised potential estimated under micro irrigation in different states varied among the states ranging from 17% in Gujarat to 67% in undivided Andhra Pradesh (Table 2.1).

**TABLE 2.1**  Potential and Actual Area (×1000 ha) under MI in Different States of India.

| State | Drip | | | Sprinkler | | | Total | | |
|---|---|---|---|---|---|---|---|---|---|
| | P* | A | % | P* | A | % | P* | A | % |
| Andhra Pradesh | 1148 | 736 | 64 | 440 | 323.6 | 74 | 1588 | 1059.6 | 67 |
| Gujarat* | 2455 | 350.3 | 14 | 1604 | 318.8 | 20 | 4059 | 669.1 | 17 |
| Karnataka | 1180 | 339.4 | 29 | 684 | 414.3 | 60 | 1864 | 753.7 | 40 |
| Maharashtra | 1241 | 835.5 | 67 | 999 | 364.8 | 36 | 2240 | 1200.3 | 50 |
| Tamil Nadu | 826 | 231.1 | 28 | 221 | 28.5 | 13 | 1047 | 259.6 | 25 |

**Note:** P = potential (Revised 2014); A = actual area. Data for Andhra Pradesh refer to undivided state. Source: [2, 8, 14, Indiastat 2010].

### 2.4.2  FARM SIZE AND AREA UNDER MICRO IRRIGATION

In all the studied states, majority of the farmers are adopting MI. The percent of area under MI to the total farm area is invariably high under small and marginal farm categories and this might be due to their limited farm size which made them to make the MI investment more effective (Table 2.2). In the case of large farmers, even though their unit cost is comparatively low due to economies of farm size, the percent area under MI is comparatively low highlighting the scope expanding the MI area under these farm categories. The cap on the MI subsidy for area might be one of the reasons for this low coverage. In the case of Gujarat, large farmers have higher coverage under MI showing the flexibility is subsidy disbursement pattern. Even though the return is high under the MI, farmers are reluctant to expand the area due to other constraints like high initial capital cost, lack of technical knowledge in the operation and maintenance of the systems, and type of crops grown. The story is same like the System of Rice Intensification (SRI) adoption where the SRI results in higher yields and income, but the adoption level is much less due to operating constraints like lack of skilled labor, high management intensity, etc. [12].

**TABLE 2.2**  Farm Size and Area Irrigated by MI Systems.

| State | Farm survey | | | | |
|-------|-------------------|------------------|----------------------------|----------------------------|-----------------------|
|       | Farmer category | % of farmers | Average farm size (ha) | Average area under MI (ha) | % of area under MI |
| Andhra Pradesh | Marginal | 6 | 0.82 | 0.72 | 87.80 |
|       | Small | 70.67 | 1.7 | 1.32 | 77.65 |
|       | Large | 23.33 | 14.08 | 3.75 | 26.63 |
| Tamil Nadu | Marginal | 13.33 | 0.62 | 0.55 | 88.71 |
|       | Small | 22 | 1.72 | 1.39 | 80.81 |
|       | Large | 64.67 | 4.67 | 3.24 | 69.38 |
| Karnataka | Marginal | 6 | 1.89 | 1.01 | 53.44 |
|       | Small | 66 | 5.71 | 1.9 | 33.27 |
|       | Large | 58 | 18.12 | 6.22 | 34.33 |
| Maharashtra | Marginal | 20 | 1.8 | 0.73 | 40.56 |
|       | Small | 16.67 | 3.75 | 2.44 | 65.07 |
|       | Large | 63.33 | 6.6 | 4.24 | 64.24 |
| Gujarat | Marginal | 2 | 0.8 | 0.52 | 65.00 |
|       | Small | 20.67 | 1.75 | 1.34 | 76.57 |
|       | Large | 77.33 | 3.65 | 3.05 | 83.56 |

**Source:** field survey: marginal = less than 1 ha; small = 1–2 ha; and large = >2 ha.

## 2.4.3  COST AND RETURNS WITH MICRO IRRIGATION

MI system cost and farmers share after subsidy varied across the farm sizes. It is comparatively lower in the larger farms compared with the other farms due to economies of scale (Table 2.3). In all the states, the quantum of actual subsidy is more than 30% which is considered less compared with the subsidy percent announced [13]. Hence this may be one of the reasons for the slow spread of the MI in different states.

**TABLE 2.3**  Micro Irrigation: Cost and Returns Across States [13].

| State | Farmer category | Average cost of drip system (Rs/ha) | Net Income (Rs/ha) | IRR (%) |
|---|---|---|---|---|
| Andhra Pradesh | M (9) | 71,380 | 15,340 | 16 |
| | S (91) | 69,794 | 17,612 | 25 |
| | L (50) | 65,373 | 17,112 | 27 |
| Tamil Nadu | M (18) | 81,302 | 16,700 | 15 |
| | S (33) | 74,509 | 15,339 | 14 |
| | L (97) | 66,908 | 23,030 | 34 |
| Karnataka | M (9) | 57,906 | 15,699 | 29 |
| | S (99) | 56,950 | 15,439 | 29 |
| | L (42) | 56,553 | 15,331 | 29 |
| Maharashtra | M (25) | 42,053 | 10,026 | 22 |
| | S (20) | 48,085 | 13,000 | 29 |
| | L (76) | 48,700 | 18,165 | 67 |
| Gujarat | M (3) | 61,795 | 14,106 | 19 |
| | S (31) | 72,482 | 19,683 | 29 |
| | L (116) | 73,195 | 19,089 | 27 |

M = marginal farmer; S = small farmer; L = large farmer; IRR = internal rate of return

**Note:** Figures in the parentheses indicate number of farmers under each farm category.

(Adapted from Palanisami, K.; Mohan, K., Kakumanu, K.; Raman, S. (2011). Spread and Economics of Micro-irrigation in India: Evidence from Nine States. Economic and Political Weekly, 2011 (June 25 – July 8), 46(26/27), 81-86. Retrieved from http://www.jstor.org/stable/23018814.)

Even though, MI could pay for the MI investment, farmers still expect the subsidy for MI because of the following reasons:

a.  MI capital intensive as it varies from Rs. 80,000 to 0.15 million per ha depending upon the crop type and type of MI systems; and farmers are reluctant to make this investment quickly;

b.  Farmers' knowledge in the operation and maintenance of the MI systems is much limited, because often the systems are facing lot of problems in terms of clogging of the filters, drippers. Also the required pressure from the pump is not always maintained due to poor conditions of the pump set resulting in low pump discharge;

c.  Except for wide spaced and commercial crops, the MI is not suitable for all crops and spacings. Except in groundwater overexploited regions, farmers in other regions do not see that MI as an immediate need. Hence, providing incentives in terms of subsidy helps the farmers to introduce the MI on their farms and save the water.

The IRR is also varying across states and farm categories, where it was ranged from 15% to 29% in case of marginal farmers, 14–29% for small farmers, and 27–67% for large farmers. The IRR is higher in case of large farmers of Maharashtra as they have a diversified and high value cropping patterns ensuring higher rate of returns. Overall, the IRR ranges from 14% to 30% across states and farm categories showing the financial feasibility of MI investment (Table 2.3).

### 2.4.4  SUGGESTIONS BY FARMERS FOR BETTER ADOPTION OF MICRO IRRIGATION SYSTEMS

Even with the proved benefits and applicability of MI systems under different farm categories, still the overall adoption level is not high. This might be due to other constraints. This chapter further examines the suggestions from farmers and also the policy recommendations at different levels. The major suggestions include:

• Provision of technical support for MI operation after installation,
• Relaxation of farm size limitation in providing MI subsidies,
• Supply of liquid fertilizers,
• Improved marketing facilities, and
• Access to more credit to expand the area under MI.

Results indicate that small farmers from Andhra Pradesh, large farmers from Tamil Nadu are in need of more technical support for the adoption and management of MI. Liquid fertilizers are highly requested in Karnataka state. Market facilities of MI systems are also important in the adoption as indicated by farmers in Tamil Nadu. Also many farmers suggested for the provision of more credit facilities to increase the area under MI (Table 2.4).

**TABLE 2.4**   Suggestions by Farmers for Better Adoption of MI Systems [13].

| State | Farmer category | Percentage of farmers opined | | | | | |
|---|---|---|---|---|---|---|---|
| | | More technical support | Supply of liquid fertilizers | Providing marketing facilities | More credit facilities for MI | No ceiling on area for subsidy | Providing training crop production |
| Andhra Pradesh | M (9) | 100.00 | 11.11 | 0.00 | 11.11 | 33.33 | 0.00 |
| | S (91) | 96.70 | 5.49 | 0.00 | 0.00 | 10.99 | 6.59 |
| | L (50) | 10.00 | 0.00 | 2.00 | 0.00 | 56.00 | 0.00 |
| Tamil Nadu | M (20) | 90.00 | 50.00 | 100.00 | 100.00 | 90.00 | 50.00 |
| | S (33) | 90.91 | 42.42 | 60.61 | 96.97 | 96.97 | 48.48 |
| | L (97) | 92.78 | 30.93 | 97.94 | 97.94 | 97.94 | 49.48 |
| Karnataka | M (9) | 11.11 | 88.89 | 11.11 | 66.67 | 0.00 | 0.00 |
| | S (99) | 5.05 | 19.19 | 22.22 | 44.44 | 5.05 | 4.04 |
| | L (42) | 4.76 | 21.43 | 23.81 | 50.00 | 59.52 | 0.00 |
| Maharashtra | M (25) | 20.00 | 24.00 | 16.00 | 32.00 | 8.00 | 88.00 |
| | S (20) | 25.00 | 30.00 | 90.00 | 100.00 | 40.00 | 70.00 |
| | L (105) | 5.71 | 21.90 | 54.29 | 53.33 | 55.24 | 50.48 |
| Gujarat | M (3) | 66.67 | 33.33 | 0.00 | 33.33 | 0.00 | 66.67 |
| | S (31) | 19.35 | 19.35 | 25.81 | 19.35 | 12.90 | 38.71 |
| | L (116) | 23.28 | 11.21 | 18.10 | 10.34 | 39.66 | 37.07 |

M = marginal; S = small; L = large farmers

Figures in the parentheses indicate number of farmers under each farm category.

(Adapted from Palanisami, K.; Mohan, K., Kakumanu, K.; Raman, S. (2011). Spread and Economics of Micro-irrigation in India: Evidence from Nine States. Economic and Political Weekly, 2011 (June 25 – July 8), 46(26/27), 81-86. Retrieved from http://www.jstor.org/stable/23018814.)

## 2.4.5   RATE OF ADOPTION OF MICRO IRRIGATION TECHNOLOGY

The rate of adoption of MI technology is still very low compared to the potential estimations, even though the states have expanded the area under MI in the recent years. The poor adoption can be attributed to number of factors such as high cost of the MI systems, irregular subsidy distribution, difficulties in using fetigation, and lack of access to credit facilities. The large farmers have the advantage of economies of scale compared with small and marginal farmers whose unit cost is comparatively high thus constraining the spread of MI by the small and marginal farmers but at the same time, the subsidy implementation mechanism constraints the large farmers interest in MI expansion as well.

Hence reducing the capital cost and increasing the technical knowhow will help the spread of the MI in a bigger way. The following cost reduction and capacity-building options are also important.

### 2.4.5.1   FIELD LEVEL

There is a good scope of reducing the system cost by slight modifications in the agro techniques to suit small and medium farms like paired row planting. Enough orientation needs to be given to the manufacturers/dealers/farmers such that most economic crop-specific design can be made. Soil texture should be one important parameter in fixing the emitter spacing. This also can reduce the system cost significantly as presently irrespective of the soil type the dripper spacing adopted is 60 cm and below. Need to redesign low-cost drip and MI systems to suit the needs of the small and marginal farmers. Periodical review of the unit cost is important as many times there is lot of time lag between the decision taken about the quantum of subsidy and the actual release. Any increase in the raw material prices during this time lag period will reflect on the actual cost of the system thus decreasing the subsidy percent at the end users' level.

### 2.4.5.2   STATE AND NATIONAL LEVEL: INDIA

Currently different government departments or agencies are involved in the implementation of the subsidy oriented MI schemes. It is important to introduce uniform subsidy norms across the regions. In this context the following subsidy disbursement and MI implementation models are suggested:

a.  Follow up with the existing but successful model. Two models viz., the Gujarat Green Revolution Company Ltd. (GGRC) and the Andhra Pradesh Micro Irrigation Project (APMIP) are seen as the best available models in terms of "capacity and quality" of implementation [10]. APMIP [1] was established as a special-purpose vehicle (SPV) housed in the Directorate of Horticulture prior to the CSS in 2003 itself [1]. GGRC [2] was established in 2005 as a SPV in the form of a public promoted company by Gujarat State Fertilizers and Chemicals Ltd., Gujarat Narmada Valley Fertilizers Company Ltd., and the Gujarat Agro Industries Corporation Ltd. Andhra Pradesh was one of the early adopters of MI, and in 2002, it had about 12% of the $5 \times 10^5$ ha under drip irrigation in India. Gujarat, at the same time, only had about 2.5% of the share. After the implementation of the improved implementation models, in Gujarat and Andhra Pradesh, the area under MI has reached $10.4 \times 10^5$ ha and $6.7 \times 10^5$ ha (as on March, 2014), respectively, in these states [2]. It is expected that the GGRC and APMIP mode of implementation will facilitate the availability of good quality equipment and access to financial resources for meeting the upfront cost of investing in the technology. However, it depends upon how each state will take these models seriously in implementing with the same spirit and commitment. Alternatively, other approaches or models that may help improve the subsidy delivery in a manner that facilitate or encourage the MI expansion are also highly warranted. A few such new models are discussed below.

b.  Incorporate the MI subsidy at the production stage itself (like fertilizers) and make the equipment available in the open market at a price lower than the current market price. The manufacturing companies can be identified and necessary incentives in terms of tax concessions, etc., can be provided. The quantum of the concessions needed for each manufacturer can be decided based on the volume of MI equipment produced, type of materials produced (accessories or main system materials), and quality of materials produced. Uniform standards can be fixed so that all the consumers can get the same quality materials at comparatively cheaper prices in the open market. However, care has to be taken in accounting the MI equipment production from the manufacturers. For example, farmers receive 90% subsidy in some states (Andhra Pradesh) and 100% in some states (Tamil Nadu) where farmers' investment is comparatively less. In such cases, small and marginal farmers may be unable to respond for the subsidized MI systems at the manufacturer level but other farmers may respond quickly in MI investment and in the long run the time lag in MI investment will be narrow down significantly.

c. Provide subsidies in the form of interest free loans by the commercial banks with no cap on the area [5]. The capital will be repaid by farmers each year @ 20%. Since MI systems will pay off returns from the year 1 onwards, farmers can easily pay the capital cost. Also farmers when they buy the MI system in the open market, it is easy to bargain and emphasize on the quality delivery and after sales maintenance warranty by the suppliers. The concerned government departments need to verify the farmers' field and give approval to the bank loans so that any misuse of loans in the name of MI systems can be avoided.

d. Provide MI subsidies directly to the farmers. This will follow the model of *Direct Benefit Transfer for Loan* (DBTL) consumers (Pahel scheme) introduced by GOI recently. In this way, unit price of the MI system components will be fixed and farmers can select their MI suppliers from the list of approved suppliers and install the MI systems in their fields by paying the full cost. Once the system installation is certified by the competent (Government department or third party inspection in terms of technical specifications, area and costs), the eligible subsidy will be credited directly to the farmers' bank account by the concerned government department in the state.

## 2.5 SUMMARY

Current level of MI adoption in India is about 14%. Demand for expanding the MI is increasing over years but the rate of adoption is low than anticipated. The present study was undertaken in five selected states and indicated higher rate of returns for MI (ranging from 14% to 67%) among different farm categories across the states. The major constraints in expanding the MI area as reported by the farmers are: higher unit cost of system, lack of technical knowhow on fertigation and time lag, and higher transaction cost in getting the MI subsidy. Key suggestions include changes in MI design and reduction in capital cost of the system and introduction of uniform and effective MI implementation models across states.

## KEYWORDS

- micro irrigation
- Government of India
- water use efficiency
- Andhra Pradesh
- Drought Prone Area Program

## REFERENCES

1. APMIP. *Andhra Pradesh Micro Irrigation Project;* Department of Horticulture: Hyderabad, 2015.
2. GGRC. Unpublished Data on MI Coverage. Personal communication, 2015.
3. GOI. PMKSY Project of Prime Minister (*Pradhan Mantri Krishi Sinchayee Yojana*); Department of Agriculture and Cooperation: Ministry of Agriculture: New Delhi, 2014.
4. INCID. *Drip Irrigation in India*; Indian National Committee on Irrigation and Drainage, Government of India: New Delhi, 1994.
5. Malik, R. P. S.; Rathore, M. S. *Accelerating Adoption of Drip Irrigation in Madhya Pradesh*; International Water Management Institute (IWMI): Colombo, 2012.
6. Molden, D. R.; Sakthivadivel, R.; Habib, Z. *Basin-Level Use and Productivity of Water: Examples from South Asia;* IWMI Research Report 49, International Water Management Institute (IWMI): Colombo, 2001.
7. Namara, R. E.; Upadhyaya, B.; Nagar, R. K. *Adoption and Impacts of Micro-irrigation Technologies: Empirical Results from Selected Localities of Maharashtra and Gujarat of India. Research Report 93;* International Water Management Institute: Colombo, 2005.
8. NCPAH. *Unpublished Draft Reports and Discussion Notes on MI Area*; National Committee on Plasticulture Application in Horticulture (NCPAH), Ministry of Agriculture, Department of Agriculture and Cooperation: New Delhi, 2014.

9. NMMI. *National Mission on Micro Irrigation: Operational Guidelines;* Government of India, Ministry of Agriculture and Cooperation: New Delhi, 2010.

10. NMMI. *Impact Evaluation Study*; Government of India, Ministry of Agriculture, Department of Agriculture and Cooperation: New Delhi, 2010.

11. NMSA. *National Mission for Sustainable Agriculture: Strategies for Meeting the Challenges of Climate Change*; Department of Agriculture and Cooperation, Ministry of Agriculture: New Delhi, 2014.

12. Palanisami, K.; Karunakaran, K.; Upali, A.; Ranganathan, C. R. Doing different things or doing it differently? Rice Intensification Practices in 13 States of India. *Economic and Political Weekly*, XLVIII(8).

13. Palanisami, K.; Mohan, K., Kakumanu, K.; Raman, S. (2011). Spread and Economics of Micro-irrigation in India: Evidence from Nine States. Economic and Political Weekly, 2011 (June 25 – July 8), 46(26/27), 81-86. Retrieved from http://www.jstor.org/stable/23018814.

14. Raman, S. State-Wise Micro Irrigation Potential in India: An Assessment. Unpublished paper, Natural Resources Management Institute: Mumbai, 2010.

15. Saleth, R. M. *Water Institutions in India: Economics, Law and Policy;* Commonwealth Publishers: New Delhi, 1996.

16. Seckler, D.; Amarasinghe, U.; Molden, D.; Radhika, D.; Barker, R. Barker. *World Water Demand and Supply, 1990 to 2025: Scenarios and Issues. Research Report 19*; International Water Management Institute: Colombo, Sri Lanka, 1998.

17. Sivanappan, R. K. Prospects of Micro Irrigation in India. *Irrig. Drain. Syst.*. **1994,** 8(1), 49–58.

18. Suresh, K. Promoting Drip Irrigation: Where and Why? Managing Water in the Face of Growing Scarcity, Inequity and Declining Returns: Exploring Fresh Approaches; *IWMI TATA 7th Annual Partner Meet*, Volume 1; 108–120, 2008.

19. Vaidyanathan, A. *Water Resources Management: Institutions and Irrigation Development in India*; Oxford University Press: New Delhi, 1999.

# PART II
# Irrigation Scheduling of Horticultural Crops

# CHAPTER 3

# DRIP AND SURFACE IRRIGATION METHODS: IRRIGATION SCHEDULING OF ONION, CAULIFLOWER, AND TOMATO

S. K. SRIVASTAVA[*]

*Vaugh School of Agricultural Engineering & Technology, SHIATS, Allahabad 211007, India*

[*]*E-mail: santoshagri.2008@rediffmail.com*

## CONTENTS

## 3.1   INTRODUCTION

Water, being the limited resource, its efficient use is essential in order to increase agricultural production per unit volume of water and per unit area of crop land. Due to increase in population, the competition of limited water resources for domestic, industrial, and agricultural needs is increasing considerably. Water for irrigation is becoming scarce and expensive due to depletion of surface and subsurface water caused by erratic rainfall and over exploitation. Therefore, it is essential to formulate economically viable water and other input management strategies in order to irrigate more land area with existing water resources and to enhance crop productivity. The improper distribution lowers the conveyance efficiency and ultimately causes the loss of irrigation water. Thus, right amount and frequency of irrigation is vital for optimum use of limited water resources for crop production and management.

The aim of irrigation scheduling is to increase efficiencies by applying the exact amount of water needed to replenish the soil moisture to the desired level. Appropriate irrigation scheduling saves water and energy. Therefore, it is important to develop irrigation scheduling techniques under prevailing climatic conditions in order to utilize scarce water resources effectively for crop production. Numerous studies have been carried out in the past in the development and evaluation of irrigation scheduling under a wide range of irrigation systems and management, soil, crop, and agro climatic conditions. The climate-based irrigation scheduling approaches (such as pan evaporation replenishment and cumulative pan evaporation and ratio of irrigation water to cumulative pan evaporation) have been used by many researchers due to simplicity, data availability, and higher degree of adaptability at the farmer's field[5].

Surface irrigation is the most common method for field/vegetable/ fruit crops and flower. The overall efficiency of surface irrigation method is considerably low compared to modern pressurized irrigation systems: drip, micro-sprinkler, and sprinkler[22, 23].

Cauliflower is a cool season vegetative crop that belongs to the species *Brassica oleracea* and family Brassicaceae. *B. oleracea* also includes cabbage, broccoli, and collard greens, though they are of different cultivar groups. The cauliflower was domesticated in north-eastern India. Cauli- flower is low in fat and carbohydrates but high in dietary fiber, folate, water, and vitamin C, possessing high-nutritional properties. Cauliflower contains several phytochemicals that are beneficial to human health.

The tomato is one of the most important vegetable crops grown in India and is regarded as a cash crop. Researchers have shown that higher tomato yield can be obtained by drip and sprinkler irrigation methods. Tomato requires warm weather and plenty of sunshine with proper irrigation for its best development and is grown well in sandy-loam to heavy-clay soils. Soil with a pH of 6–7 is ideal. The tomato crop occupies an area of 36,682 ha producing about 714,200 tons annually with productivity of 21.20 tons/ha, in Uttar Pradesh. The local survey indicates that till today farmers use gravity irrigation methods that result in wastage of water due to heavy conveyance and water application losses.

Onion is a shallow rooted (8–15 cm) crop, and is sensitive to irrigation and requires frequent watering as compared to other vegetable crops. Among various factors for higher yield, the use of appropriate quantity of water and nutrients at proper time are important, hence desired yield gain is never achieved due to inadequate management of water/nutrients in the soil under limited water resource conditions to fulfil the crop's water requirements. Efficient irrigation strategies are essential in order to maximize yield. Due to lack of knowledge of efficient irrigation practices, the average yield and irrigation production efficiency (IPE) of onion is somewhat low[10].

Onion roots are shallow and not very efficient in taking up moisture. Therefore, they need a steady supply of water to grow without interruption. Although, they actually recover well from drought and start growing again when watered, it is best to keep the soil consistently moist until the bulbs are enlarged.

Therefore, the present study is undertaken with following objectives to improve marketable yield[13], water-use efficiency, and economic return of onion, cauliflower, and tomato under drip and surface irrigation methods:

a.  To investigate the effects of irrigation scheduling on yield and IPE of onion, cauliflower, and tomato.

b.  To investigate the effects of irrigation scheduling on economic returns of onion, cauliflower, and tomato.

c.  To develop water-yield relationships of onion, cauliflower, and tomato in order to optimize the yield under limiting water supply conditions.

d.  To develop relationships between seasonal water application/ irrigation scheduling and crop yield, gross return, net return, and benefit cost ratio (BCR) in order to maximize the profits: onion, cauliflower, and tomato.

## 3.2   REVIEW OF LITERATURE

### 3.2.1   IRRIGATION METHODS

The irrigation is a method of applying water to the land surface. Irrigation water may be applied to crops by flooding it on the field surface, by applying it beneath the ground surface (sub-surface irrigation), by spraying it under pressure (sprinkler) or by applying it in drops near the root zone (drip or trickle irrigation). The quantity and quality of water available, the topography of the land, the crop to be irrigated, the cost of the water application system and the availability of labor will determine the most desirable method of irrigation. The efficient irrigation results in increased crop yields, good soil fertility maintenance, and economical water use.

Overirrigation, however, results in leaching of fertilizers, water logging, and salt accumulation. Excess water application rate from streams or surface irrigation on sloping lands result in soil erosion. Whatever be the method of water application, it is essential that the system is designed to apply the right amount of water at the right time and apply it uniformly to meet demand of soil moisture near the root zone.

Surface irrigation is the most common and traditional method throughout the world. Due to increasing water scarcity caused by erratic rainfall, over exploitation of water resources, drip irrigation is considered to have advantages over other types of irrigation systems.

It is reported that drip irrigation has greater water saving over other system in arid and semi-arid regions characterized by high evaporation rates. It has the potential to increase crop yield even with reduced irrigation water application. Also drip irrigation can help to irrigate hilly terrains or texturally nonuniform fields. It also allows farming on flat lands to save labor and operating cost of land levelling, making furrows, and ridges[25–28].

Numerous studies have been carried out to compare the methods of irrigation and its response to the vegetable crops and orchards, and significant results were found in different agro-climatic zones, and soil conditions[14, 15]. Drip irrigation reduces soil evaporation and deep percolation, controls soil moisture more precisely, and reduces the effects of wind[1]. Bucks et al.[6] found similar cabbage yields for furrow and drip irrigation for adequately watered conditions. Higher yield of fresh market tomatoes under drip irrigation was recorded than furrow irrigation[6]. Curmen et al.[9] record higher potato yield under sub-surface irrigation than drip, sprinkler, and furrow methods. Bucks et al.[6] reported that drip irrigation offers numerous advantages compared to furrow irrigation; 44% of water saving was obtained; and runoff was minimum.

## 3.2.2  IRRIGATION SCHEDULING

Irrigation scheduling includes when and how much water to apply to the crop. It maximizes irrigation efficiency due to application of exact amount of water needed to replenish the soil moisture to the desired level. Irrigation scheduling saves water and energy. All irrigation scheduling methods consist of monitoring and determining the need for irrigation.

Irrigation scheduling is a critical management input to ensure optimum soil moisture status for proper plant growth and development, optimum yield, and fruit quality, water use efficiency, and economic benefits. Irrigation scheduling to determine the frequency and amount of water application is governed by various factors, but climate plays an important role. Therefore, it is essential to develop irrigation scheduling strategy under local climate conditions to utilize water source more effectively and efficiently. Irrigation provides means for optimum plant water use and crop yield. Implementing sound irrigation water management practices is necessary to overcome excessive irrigation and eliminate many associated problems. Irrigation scheduling becomes crucial element in reducing deep percolation and improving water quality downstream[1]. Soil moisture condition affects nutrient availability to crop. The relationship between yield and crop water use has been investigated and studies have been carried out also on development and evaluation of irrigation scheduling techniques under the wide range of irrigation system and management, soil, crop, and climatic conditions[12, 24].

Choudhary and others[3, 8, 18, 28] have reported that when the same amount of water was applied at different growth stages there was significant difference in the productive phase. Imtiyaz et al.[13] reported higher marketable yield of cabbage, carrot, and onion with irrigation at 22–25% of available soil moisture depletion, whereas irrigation at 42–45% soil moisture depletion resulted in higher water-use efficiency of cabbage, carrot, and onion. Imtiyaz et al.[11, 4] observed higher yield and IPE at 120% of pan evaporation replenishment of green mealies[11] under drip irrigation system.

### 3.2.3 PLANT-BASED METHODS OF IRRIGATION SCHEDULING

Plant-based methods of irrigation scheduling define measurable plant parameter, which can be used to determine when to apply irrigation water. The use of infrared thermometry helps to determine canopy water stress index (CSWI) as the basis for scheduling. It is based on the assumption that as water becomes limiting, transpiration is reduced, and plant temperature rises[2]. The significant difference in seasonal water application depth was observed under four methods of irrigation scheduling (40% depletion of root zone available water, scheduling based on plant temperature: canopy water stress index (CSWI) of 0.4, soil metric potential (SMP) of 30 kPa[3, 26]. Curmen et al.[9] suggested soil moisture potential (SMP) of −25 kPa and once a day drip irrigation frequency for potato under drip irrigation.

### 3.2.4 METEOROLOGICAL APPROACHES TO IRRIGATION SCHEDULING

Pan evaporation is the most common approach for irrigation scheduling[7]. Many researchers have used the pan evaporation replenishment, crop evaporation, and the ratio of irrigation water depth (IW) and cumulative pan evaporation (CPE) for irrigation scheduling due to its simplicity, data availability an easy accessibility at the farmer's level. Jensen et al.[16] estimated crop water use (ET) using the modified penman equation to generate crop coefficients from planting to full cover and days after full cover. Singh et al.[25] and Imtiyaz et al.[12 to 15, 19] investigated effects of scheduling using the pan evaporation on performance of crops.

Researchers found no significance differences in corn yield between the 1.0 ET and 1.33 ET treatments. Singh and Mohan[24] reported a reduction of sugarcane yield when irrigation was applied beyond IW/CPE ratio of 1.0. Sammis et al.[22, 23] conducted research on tomatoes grown on fine sandy with black polythene mulch and irrigated with drip and found higher yield at 1.0 and 0.75 pans; and water use was higher at 0.75 pan scheduling.

### 3.2.5 WATER PRODUCTION FUNCTIONS

Crop water production function is a relationship between water application depth/water use and crop yield. The information on the crop water production function is important to assess the properties for allocating the limited irrigation water. Crop water production functions are required for water management and the design of irrigation methods. Water management variables such as irrigation frequency, time of irrigation, and water allocation are important in the design of irrigation systems.

Researchers found linear relationships between soil moisture tension and yield of alfalfa and sugar beets; and curvilinear relationship for potatoes[9]. Zang et al.[29] found a linear relationship between water transpired and dry matter yield. Mishra et al.[20] reported field data for wheat under drip irrigation.

Thus for water resource and management, water crop production functions can play an important role in both production decisions and policy analysis. A production function, which mathematically or graphically represents the relationship between inputs and outputs in a production process, serves as a basis for describing and predicting the expected output from a specified level of inputs. Production function for irrigated agricultural crops can be determined directly from experiments, from statistical analysis of secondary data, or intricately by mathematical simulation models. Simulation models can be readily adapted to specific soil and climatic conditions, and to provide a flexible and relatively inexpensive method of producing production functions for varying local condition.

The conventional relationship between water application depth and crop yield, water crop production function simulation can provide estimates of another policy relevant variable, the consumptive use or evapotranspiration associated with a given irrigation scheduling.

### 3.2.6 ECONOMIC RETURNS

The initial investment of drip irrigation system is higher as compared to surface irrigation. Government officials and policy makers have stated that drip irrigation system is economical for vegetable production in India[6]. Numerous studies in the past have reported that drip irrigation has resulted in greater economy compared to other systems and gives higher profits for vegetables and fruits. It has also the potential to increase crop yield even with reduced irrigation application depth. The amount of water used for irrigation is constantly being assessed because of concerns about rising energy costs, deep percolation of water and dissolved chemicals, increased competition of water use among various crops, and declining ground water supplies. The total cost of production under any irrigation system includes fixed and operating costs and capital investment cost. Capital investment cost is the investment in basic irrigation infrastructure and equipments without which production could not take place. These include water supply system, irrigation system layout, and automatic water control. Fixed annual costs are expenditures incurred as a result of occurring capital investment; and these include depreciation, interest, taxes, and repairs. Variable costs include electricity for pumping water and labor for cleaning and flushing drip lines. The operating cost includes labor, land, seeds, fertilizer, chemicals, and repair and maintenance.

Production practices and physical conditions also have an effect on the economic returns. Early plant harvest dates can result in less water for irrigation because of lower evapotranspiration rates early in the season. Depending on the market conditions[13], early harvest dates can also mean higher prices of products. Physical factors like rainfall and soil type also affect the financial appeal. The system also plays a crucial role in the financial appeal of the irrigation system. The emitters are more economical than those with extruded emitters, especially when the system was used for several seasons. It has also been found that for a single-season use, the biwall pipe system and spiral-in-line emitter system are more useful.

Imtiyaz et al.[11] conducted field experiment to study the effects of three levels of pan evaporation replenishment (20, 40, 60%) on economic return of winter broccoli, carrot, rape, and cabbage under drip irrigation method[28]. The results indicated that the net return was increased with increase in pan evaporation replenishment. The maximum net return (Rs./ha) were 71,600 for broccoli, 59,400 for carrot, 81,200 for rape, and 1900 for cabbage, respectively, with irrigation at 80% and 100% of pan replenishment. Imtiyaz et al.[15] studied effects of irrigation scheduling (18 mm water depth in each irrigation at 11, 22, 33, 44, and 55 mm of cumulative pan evaporation) on economic returns of cabbage, spinach, rape, carrot, tomato, and onion under drip irrigation.

## 3.3   METHODS AND MATERIALS

Field experiment was conducted to study the effects of drip and surface irrigation systems and irrigation management on performance of onion, cauliflower, and tomato. The necessary data on marketable yield, IPE, and economic returns were collected.

Field experiment was conducted at the irrigation research farm of Sam Higginbottom Institute of Agriculture, Technology and Sciences at Allahabad, India, (25°27′N, 81°44′E, 98 m above mean sea level) during the winter growing season of November 2013 to March 2014 for onion, October 2013 to March 2014 for cauliflower. The soil at the site was fertile clay loam. The climate at the location has been classified as semi-arid with cold winter and hot summer. The average temperature during experimental period ranged between 21°C and 40°C. Rainfall during crop growing season is recorded 1.21 mm. The climatic parameters during crop growing period are presented in Table 3.1.

**TABLE 3.1** Climatic Parameters during Winter Crop Growing Season.
(November 2013–April 2014)

| Months | Mean wind velocity, (km/h) | Mean sunshine (h) | Mean humidity (%) | Mean maximum temperature (°C) | Mean minimum temperature (°C) | Mean evaporation (mm) |
|---|---|---|---|---|---|---|
| Nov. | 0.68 | 8.25 | 69.03 | 29.53 | 15.10 | 3.78 |
| Dec. | 1.29 | 6.83 | 69.40 | 27.20 | 13.24 | 2.88 |
| Jan. | 1.63 | 3.67 | 76.34 | 21.82 | 11.66 | 2.21 |
| Feb. | 1.49 | 6.51 | 69.61 | 25.76 | 13.32 | 2.35 |
| Mar. | 1.27 | 7.92 | 62.71 | 32.20 | 18.19 | 2.89 |
| Apr. | 1.59 | 8.63 | 51.62 | 38.93 | 21.89 | 3.95 |

### 3.3.1 EXPERIMENTAL LAYOUT

The total experimental area was about 280 m², which was well irrigated, properly ploughed, and well pulverized and levelled.

The whole area was divided into 24 plots each of 9.0 m² leaving a 0.5 buffer zone between the plots according to the experimental layout; drip, and surface irrigation systems were installed in the respective plots. The experimental block design was used with three replications, four irrigation levels, and two irrigation methods. In total, there were eight treatments. Layout of the experimental plots is shown in Figure 3.1.

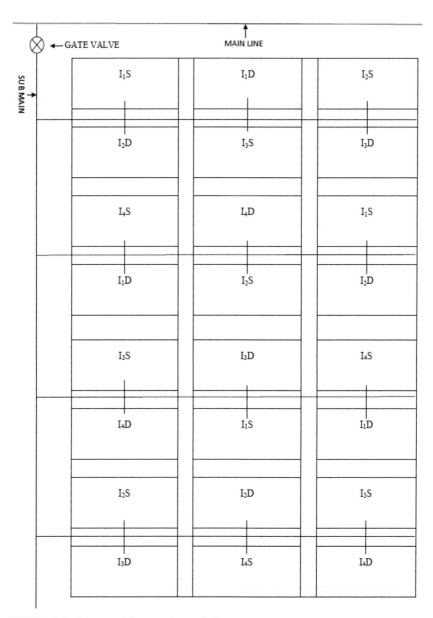

**FIGURE 3.1**    Layout of the experimental plots.

### 3.3.2   TREATMENTS

The experiments consisted of four irrigation levels and two irrigation methods. The details of treatments are given below: treatments = irrigation level × irrigation method × replications.

### A. Irrigation levels

$I_1$ = Irrigation at 25% of pan evaporation replenishment.
$I_2$ = Irrigation at 75% of pan evaporation replenishment.
$I_3$ = Irrigation at 125% of pan evaporation replenishment.
$I_4$ = Irrigation at 175% of pan evaporation replenishment.

### B. Irrigation methods

S = Check basin (surface irrigation)
D = Drip irrigation

### C. Surface irrigation method

Surface irrigation method, also called gravity irrigation, comprised of water application in which water was distributed by means of open-surface flow. Two basic requirements of prime importance to obtain high efficiency in surface irrigation are properly constructed water distribution systems to provide adequate control of water to the field and proper land preparation to obtain uniform distribution of water over the field. Water was applied to the land surface from a field supply channel located at the upper end of the field and water flowed into the field under gravity. In case of check basin method, water was applied through pipe conveyance system. The harvesting was done manually.

### D. Drip irrigation method

Drip irrigation is the method of localized slow application of water to the plant root zone[17]. The system uses pipe lines, tubes, filters, emitters, and ancillary devices to deliver water to specific sites at a point or grid on the soil surface. The water oozes in drops from the emitters and hence the name drip irrigation. In the drip system, losses by deep percolation and evaporation are minimized. Precise amount of water is applied to replenish the depleted soil moisture at frequent intervals, for optimum plant growth. The system enables the application of water and fertilizer at an optimum rate to the plant-root system. The amount of water supplied to the soil can be adjusted to equal the daily consumptive use, thus maintaining a low moisture tension in the soil.

The crop was irrigated either by surface and drip irrigation. The screen filters were installed to minimize clogging of drippers and sprinklers. PVC pipes of 50 mm diameter and LDPE of 12 mm diameter were used for main/sub-main and lateral lines, respectively. The harvesting was done manually.

### 3.3.3  IRRIGATION PRODUCTION EFFICIENCY

An irrigation management practice is evaluated to access better utilization of irrigation water. The evaluation is expressed in terms of crop yield per unit volume of water applied. Irrigation production efficiency is determined as follows:

$$\text{Irrigation production efficiency} = \frac{[\text{crop yield, kg/ha}]}{[\text{seasonal water depth, m}^3/\text{ha}]} \qquad (3.1)$$

### 3.3.4  ECONOMIC ANALYSIS

In order to assess the economic viability of drip and surface irrigation systems under variable irrigation depths, both fixed and operating cost were estimated. Total cost of production, gross return, net return, and BCR were estimated in this study. The fixed cost included cost of water development, irrigation equipments, spraying, and weeding. Following assumptions were made:

Total cost of production, gross return, net return, and BCR under variable irrigation is estimated as per the following assumption:

- Salvage value = 0
- Useful life of tube well, pump, motor, and pump house = 25 years
- Useful life of weeding and spraying equipments = 7 years
- Useful life of drip irrigation system = 8 years
- Useful life of surface irrigation system = 5 years
- Interest rate = 12.5%
- Repair and maintenance = 7.5%
- Number of crops/year = 2

$$\text{Capital Recovery Factor, CRF} = i(1 + i)^n/(1 + i)^n - 1 \qquad (3.2)$$

$$\text{Annual fixed cost/ha} = \text{CRF} \times \text{fixed cost/ha} \qquad (3.3)$$

$$\text{Annual fixed cost/ha/season} = (\text{annual fixed cost/ha})/2 \qquad (3.4)$$

Where, $i$ = interest rate (fractions); and $n$ = useful life of the component (years).

$$\text{Total cost of production} = \text{fixed cost} + \text{operating cost} \qquad (3.5)$$

$$\begin{aligned} \text{Gross return (Rs./ha)} = \text{marketable yield (t/ha)} \times \\ \text{wholesale price of onion (Rs./t)} \end{aligned} \qquad (3.6)$$

$$\begin{aligned} \text{Net return (Rs./ha)} = \text{gross return (Rs./ha)} \\ - \text{total cost of production (Rs./ha)} \end{aligned} \qquad (3.7)$$

$$\begin{aligned} \text{Benefit cost ratio} = \text{gross return (Rs./ha)}/ \\ \text{total cost of production (Rs./ha)} \end{aligned} \qquad (3.8)$$

The operating cost included cost for labor (system installation, irrigation, planting, weeding, cultivation, fertilizers, and chemical application, harvesting, packing, etc.) land preparation, land rent, seeds, fertilizers, chemicals, water pumping, and repair and maintenance (tube well, pump, electric motor, pump house, water tank, irrigation systems, etc.). The gross return was calculated taking into consideration the yield and current wholesale price of onion. Subsequently, the net return for onion was calculated considering total cost of onion production and gross return.

### 3.3.5  COLLECTION OF PAN EVAPORATION DATA

The daily pan evaporation data at the site was collected from Agro-meteorological Department at AAI, SHIATS, Allahabad, and daily mean pan evaporation for the crop growing season was estimated. Crop was irrigated when the sum of daily mean (for 5 years: 2008–2012) USWB Class A pan evaporation reached the desired value. It was estimated from the soil moisture, plant available water and readily available water.

### 3.3.6  CROP ESTABLISHMENT, CULTIVATION, AND PRODUCTION

Onion (Red Nasik-Hybrid N-53) seed was sown in the nursery on 16 November 2014, at a depth of 0.05 m with a spacing of 10 cm between the rows. Onion seedlings were then transplanted into plots at spacing of plant-to-plant 15 cm and row-to-row 30 cm.

Cauliflower (US-5012, F1-Hybrid) was sown in the nursery at a depth of 0.05 m. Cauliflower seedlings were then transplanted into plots at spacing of 0.5 m × 0.5 m (row-to-plant). First irrigation was applied just after transplanting. The picking of cauliflower heads was done from 21st Feb to 15th march. The total number of cauliflower heads and total yield was obtained by weighing the crop for each treatment separately and it was considered as a marketable yield. Average mean weight of cauliflower was calculated accordingly for each treatment.

Tomato (var. F1-Hybrid SHUBHAM-0905) seed was sown on 16 November 2013, in the nursery at a depth of 0.05 m with a spacing of 0.10 m between the rows. The Seedlings were transplanted on 19 December 2013, at a spacing of 0.5 m × 0.5 m. The crop was harvested manually during 3–23 April 2014, depending upon the maturity of fruits.

### 3.4  RESULTS AND DISCUSSION

### 3.4.1  ONION

This section of research study on onion was conducted by Abhay Bara and Suman Tigga, graduate students under the supervision of Professor S. K. Srivastava.

## 3.4.1.1   YIELD AND IRRIGATION PRODUCTION EFFICIENCY

Yield and IPE of onion influenced by irrigation scheduling and irrigation methods are presented in Table 3.2. Appendix A1 shows the effect of irrigation scheduling levels and irrigation methods on seasonal water application and marketable yield of onion. Appendices B1 and B2 show the statistical analysis (ANOVA Tables) for effects of irrigation scheduling and irrigation methods on seasonal water application, IPE and marketable yield of onion. The mean crop yield for different irrigation levels ranged from 15.3 to 24.6 t/ha. Irrigation at 125% of pan evaporation replenishment resulted in significantly higher mean crop yield of 27.55 t/ha. A further increase in irrigation level resulting from 175% of pan evaporation replenishment reduced the mean crop yield (24.6 t/ha) significantly. Irrigation at 25% of pan evaporation replenishment resulted in significantly minimum crop yield (15.3 t/ha). Irrigation methods significantly influenced the mean crop yield of onion.

Irrigation levels and irrigation methods had marked effect on IPE of onion (Table 3.2). The IPE for different irrigation levels ranged from 3.62 to 15.77 kg/m$^3$. The IPE was decreased significantly with the increase in irrigation levels because increase in the mean crop yield was lower than the seasonal water application. Irrigation at 25% of pan evaporation replenishment resulted in higher mean IPE of 15.77 kg/m$^3$, because reduction in seasonal water application was higher than the reduction in crop yield.

A further increase in irrigation levels from 25 to 175% of pan evaporation replenishment reduced the IPE significantly, because increase in crop yield was less than the increase in seasonal water application. Irrigation at 175% of pan evaporation replenishment resulted in significantly minimum IPE of 3.62 kg/m$^3$, because it increases the seasonal water application considerably but the crop yield was decreased. Irrigation methods significantly influenced the IPE of onion (Table 3.2). The irrigation efficiency of onion for drip and surface irrigation methods was 9.49 and 9.71 kg/m$^3$, respectively.

Drip irrigation method resulted significantly higher and IPE, due to higher crop yield. Surface irrigation resulted in minimum IPE, due to considerably low mean crop yield. The overall results in Table 3.2 revealed that both irrigations levels and methods influenced the mean crop yield and mean IPE of onion. The highest mean crop was recorded when irrigation during the crop growing season was applied at 125% of pan evaporation replenishment, whereas mean IPE was higher with irrigation level at 25% of pan evaporation replenishment.

**TABLE 3.2**   Effects of Irrigation Scheduling and Irrigation Methods on Marketable Yield, Yield Components, and Irrigation Production Efficiency of Onion.

| Treatment | Mean yield (t/ha) | Mean irrigation production efficiency (kg/m³) |
|---|---|---|
| **Irrigation scheduling (pan evaporation replenishment, %)** | | |
| 25 | 15.3 | 15.77 |
| 75 | 21.65 | 13.34 |
| 125 | 27.55 | 5.67 |
| 175 | 24.6 | 3.62 |
| CD (0.05) | 1.121 | 0.800 |
| Irrigation methods | | |
| Drip | 26.65 | 9.49 |
| Surface | 17.9 | 9.71 |
| CD (0.05) | 0.793 | 0.565 |
| **Interaction (irrigation scheduling × methods)** | | |
| CD (0.05) | 1.585 | 1.131 |

The seasonal water application depth ranged from 97 to 679 mm whereas the IPE of onion varied from 17.2 to 29.9 and 13.4 to 19.3 kg/m³ for drip and surface irrigation methods (Table 3.2). The seasonal water application and IPE of onion for drip ($R^2 = 0.971$) and surface ($R^2 = 0.9222$) irrigation methods exhibited a linear relationship. The IPE was decreased with the increase in seasonal water application.

The pan evaporation replenishment ranged from 25 to 175% whereas the IPE of onion for ranged from 17.2 to 29.9 kg/m³ for drip and 13.4 to 19.3 kg/m³ for surface irrigation, respectively (Table 3.2). The pan evaporation replenishment and IPE of onion for drip ($R^2 = 0.971$) and surface ($R^2 = 0.922$) irrigation methods exhibited a linear relationship. The IPE was decreased with increase in pan evaporation replenishment. The results revealed that higher pan evaporation replenishment and seasonal water application did not increase evapotranspiration as well as onion yield of onion but it increased the deep percolation. In spite of some variation, overall results show the quadratic relationship between crop yield and seasonal water application, and pan evaporation replenishment for two irrigation methods. Irrigation production efficiency and seasonal water application for two irrigation method exhibited a linear relationship.

### 3.4.1.2   WATER APPLICATION DEPTH AND MARKETABLE YIELD

The relationships between seasonal water application and marketable yield of onion[13] are represented in Figure 3.2, for both irrigation methods. The seasonal water application varied from 97 to 679 mm, whereas crop yield ranged from 17.2 to 29.4 t/ha for drip and 13.4 to 19.3 t/ha for surface irrigation, respectively. The seasonal water application and crop yield of onion for drip ($R^2 = 0.971$) and surface ($R^2 = 0.922$) irrigation methods exhibited strong quadratic relationships. The results revealed that higher seasonal water application did not increase the evapotranspiration as well as the

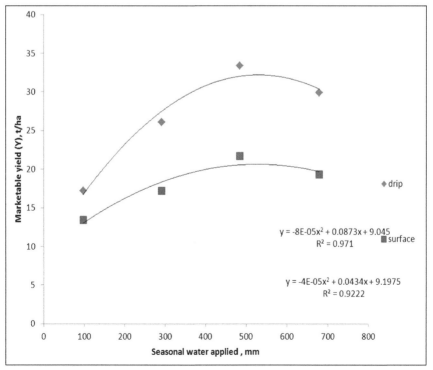

**FIGURE 3.2**   Relationship between seasonal water application and onion yield for different irrigation methods (top: drip, $R^2 = 0.971$; bottom: surface, $R^2 = 0.922$).

crop yield but it increased the deep percolation. The quadratic yield and seasonal water application relationships probably resulted from nutrient leaching through deep percolation and poor soil aeration (Table 3.2). The pan evaporation replenishment ranged from 25 to 175% whereas the onion yield for drip and surface irrigation methods ranged from 15.3 to 24.6 t/ha, and 15.77 to 3.62 t/ha, respectively.

The pan evaporation replenishment and the onion yield for drip ($R^2 = 0.971$) and surface ($R^2 = 0.9222$) irrigation methods exhibited strong quadratic relationships (Fig. 3.3). The crop yield was increased with increase in pan evaporation replenishment and attained its maximum value for drip and surface irrigation methods at 125% of pan evaporation replenishment and there after it started to decline. The quadratic relationships between water application and onion yield for different irrigation methods were probably due to poor aeration and nutrient leaching caused by excessive soil moisture.

### 3.4.1.3   ECONOMIC RETURNS

Table 3.3 presents total cost of production, gross returns, net returns and BCR of onion under two irrigation methods (drip and surface) and irrigation scheduling levels. Appendices A2–A7 show the raw data and estimations for economic returns of onion production under two irrigation methods. The total cost of production was increased slightly with increase in irrigation levels due to insignificant increase in pumping cost induced by variation in seasonal water applied (Appendices A2–A7).

**TABLE 3.3**   Economic Returns of Onion under Different Irrigation Scheduling and Irrigation Methods.

| Treatment (pan evaporation replenishment) (%) | Total cost of production (Rs./ha) | | Gross return (Rs./ha) | | Net return (Rs./ha) | | Benefit cost ratio (BCR) | |
|---|---|---|---|---|---|---|---|---|
| | Drip | Surface | Drip | Surface | Drip | Surface | Drip | Surface |
| 25 | 81,087 | 46,634 | 309,600 | 241,200 | 228,513 | 194,566 | 3.81 | 5.1 |
| 75 | 83,842 | 49,389 | 446,980 | 309,600 | 385,958 | 260,211 | 5.6 | 6.25 |
| 125 | 86,596 | 52,144 | 601,200 | 390,600 | 51,4603 | 338,436 | 6.94 | 7.49 |
| 175 | 89,351 | 54,899 | 538,300 | 347,500 | 448,848 | 292,502 | 6.02 | 6.42 |

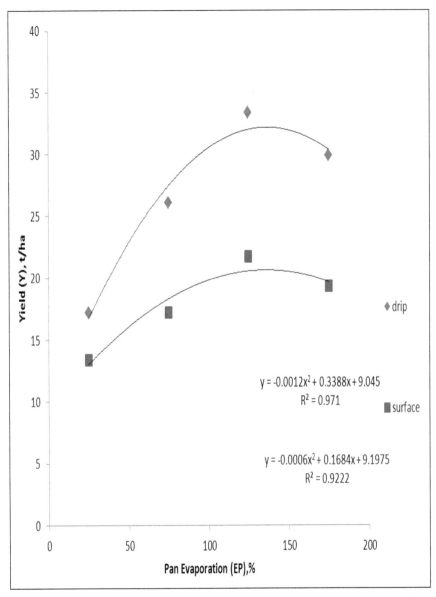

**FIGURE 3.3** Relationship between pan evaporation replenishment and onion yield for two irrigation methods (top: drip, $R^2 = 0.971$; bottom: surface, $R^2 = 0.922$).

The cost of production on onion varied from 81,087 to 89,352 Rs./ha for drip and 46,634 to 54,899 Rs./ha for surface irrigation. The total cost of production in drip irrigation was considerably higher as compared with surface irrigation methods mainly due to variation in cost of drip irrigation system cost due to large number of drippers/ha and lateral length. The gross return of onion under different irrigation levels ranged from 309,600 to 601,200 Rs./ha for drip and 241,200 to 390,600 Rs./ha for surface irrigation. The gross return was increased sharply from 25 to 175% pan evaporation replenishment due to increase in crop yield. A further increase in pan evaporation replenishment decreased the gross return due to decrease in crop yield.

The net return of onion under different irrigation levels ranged from 228,513 to 514,603 Rs./ha for drip and 194,566 to 338,456 Rs./ha for surface irrigation. The net return was increased sharply from 25 to 175% of pan evaporation replenishment due to increase in crop yield. A further increase in the irrigation level at 175% of pan evaporation replenishment reduced the net return. The maximum return of drip (514,603 Rs./ha) and surface (338,456 Rs./ha) were obtained when irrigation level during the crop growing season was applied at 125% of pan evaporation replenishment (Table 3.3).

The BCR under different irrigation levels ranged from 3.87 to 6.94 for drip and 5.1 to 7.49 for surface irrigation. The BCR of onion for two irrigation methods was increased considerably from 25 to 175% of pan evaporation replenishment due to sharp increase in gross return. A further increase in pan evaporation replenishment irrigation levels decreased the BCR. The maximum BCR of onion for drip (6.94) and surface irrigation (7.49) methods were obtained at 125% of pan evaporation replenishment.

### 3.4.1.4   WATER APPLICATION DEPTHS AND ECONOMIC RETURNS

The relationships between seasonal water application depth and gross return of onion under two irrigation methods are presented in Figure 3.4. For seasonal water application depth from 97 to 679 mm, the gross return ranged from 309,600 to 601,200 Rs./ha for drip and 241,200 to 390,600 Rs./ha for surface irrigation. Seasonal water application depth and gross return of onion for drip ($R^2 = 0.971$) and surface ($R^2 = 0.922$) irrigation methods

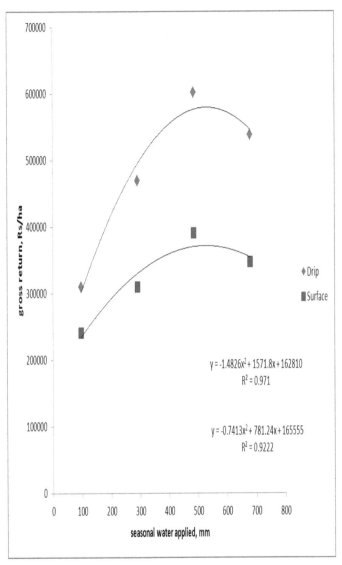

**FIGURE 3.4**  Relationship between seasonal water application and gross return of onion for two irrigation methods (top: drip, $R^2 = 0.971$; bottom: surface, $R^2 = 0.922$).

exhibited strong quadratic relationships. The results revealed that fitted regression models can be used for optimizing gross return of onion under different irrigation levels and irrigation methods.

The relationships between pan evaporation replenishment and gross return of onion for different irrigation methods are shown in Figure 3.5. For pan evaporation replenishment of 25–175%, the gross return of onion for drip and surface irrigation methods ranged from 309,600 to 601,200 Rs./ha for drip and 241,200 to 390,600 Rs./ha for surface irrigation. The pan evaporation replenishment and gross return for drip ($R^2 = 0.971$) and surface ($R^2 = 0.922$) irrigation methods exhibited strong quadratic relationships. The results revealed that fitted regression models can be used for optimizing gross return of onion under different irrigation levels and irrigation methods.

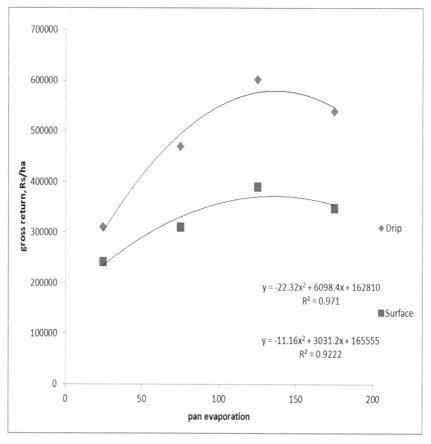

**FIGURE 3.5** Relationship between pan evaporation replenishment and gross return of onion for different irrigation methods (top: drip, $R^2 = 0.971$; bottom: surface, $R^2 = 0.922$).

The relationships between seasonal water application depth and net return of onion for drip and surface irrigation methods are illustrated in Figure 3.6. The relationships between pan evaporation replenishment and net return of onion for drip and surface irrigation methods are illustrated in Figure 3.7. The relationships between seasonal water application depth and BCR of onion for drip and surface irrigation methods are illustrated in Figure 3.8. The relationships between pan evaporation replenishment

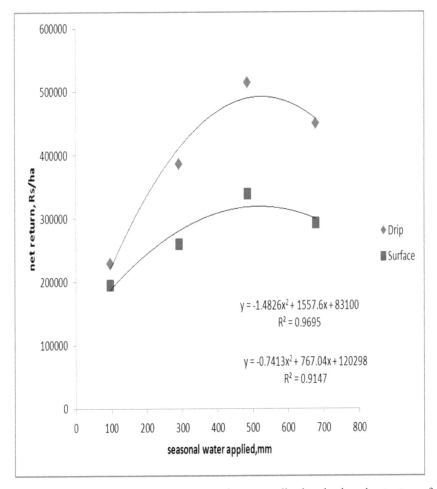

**FIGURE 3.6**  Relationship between seasonal water application depth and net return of onion for different irrigation methods (top: drip, $R^2 = 0.9695$; bottom: surface, $R^2 = 0.9147$).

and BCR of onion for drip and surface irrigation methods are illustrated in Figure 3.9. For seasonal water application depth of 97–679 mm, the net return of onion ranged from 228,513 to 514,603 Rs./ha for drip and 194,566 to 338,456 Rs./ha for surface irrigation. Despite some variation, the seasonal water application and net return gross of onion for drip

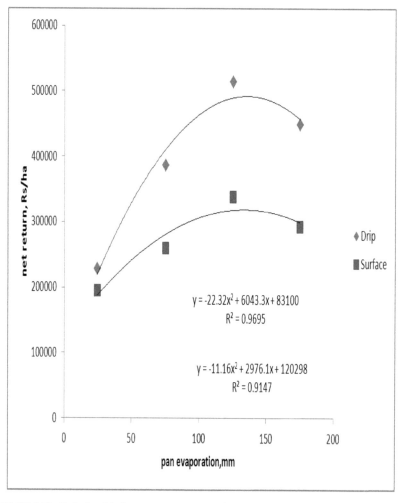

**FIGURE 3.7**    Relationship between pan evaporation replenishment and net return of onion for different irrigation methods (top: drip, $R^2 = 0.9695$; bottom: surface, $R^2 = 0.9147$).

($R^2$ = 0.9695) and surface ($R^2$ = 0.9147) irrigation methods exhibited strong quadratic relationships (Fig. 3.6). The results in Figures 3.4–3.9 revealed that fitted regression models can be used for optimizing net return gross of onion under different irrigation levels and irrigation methods.

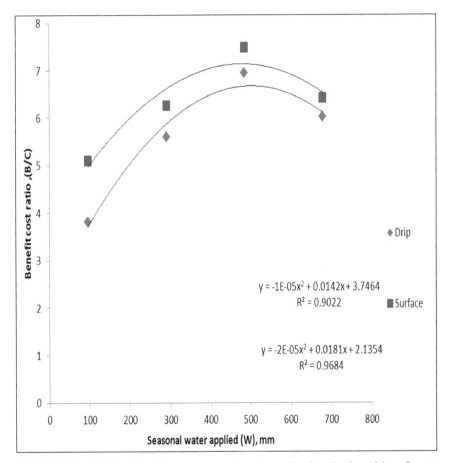

**FIGURE 3.8** Relationship between seasonal water application depth and benefit cost ratio of onion for different irrigation methods (top: drip, $R^2$ = 0.9684; bottom: surface, $R^2$ = 0.9022).

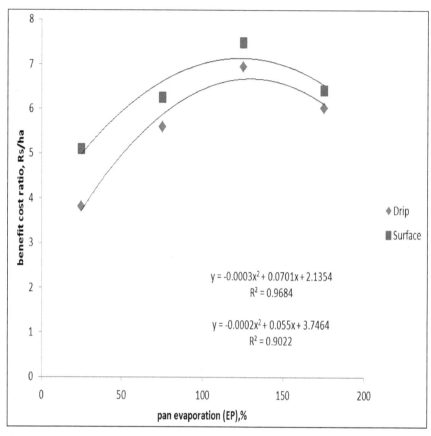

**FIGURE 3.9** Relationship between pan evaporation replenishment and benefit cost ratio of onion for different irrigation methods (top: drip, $R^2 = 0.9684$; bottom: surface, $R^2 = 0.9022$).

### 3.4.1.5 CONCLUSIONS

The experimental results showed that irrigation at 125% of pan evaporation replenishment gave significantly maximum yield of onion due to higher crop weight, whereas irrigation at 25% of pan evaporation replenishment resulted in highest IPE.

Drip irrigation method resulted in higher crop yield and IPE followed by surface irrigation method. Although drip irrigation method is costly, yet the results clearly depict that higher profits were resulted from high production irrigation efficiency.

Finally, the overall results clearly suggest that in order to obtain higher crop yield, IPE and net return of onion, during the winter growing season (December–April), the onion crop should be irrigated at 125% of pan evaporation replenishment under drip or surface irrigation method.

### 3.4.2   CAULIFLOWER

This section of the research study on cauliflower was conducted by Pratima Horo, Shailendra Tirkey, and Geoffrey Bai Passah, graduate students under the supervision of Professor S. K. Srivastava.

#### 3.4.2.1   MARKETABLE YIELD AND IRRIGATION PRODUCTION EFFICIENCY

Cauliflower marketable yield and IPE of cauliflower were influenced by irrigation levels and irrigation methods (Table 3.4). The mean crop yield for different irrigation levels ranged from 17.05 to 37.66 t/ha. Irrigation at 125% of pan evaporation replenishment resulted in significantly higher mean crop yield (37.66 t/ha). A further increase in irrigation level to 175% of pan evaporation replenishment resulted in significantly minimum crop yield (32.61 t/ha). Irrigation at 25% of pan evaporation replenishment resulted in significantly minimum crop yield (17.05 t/ha). Irrigation methods significantly influenced the mean crop yield of cauliflower. The drip irrigation gave significantly higher yield (33.1 t/ha) compared to surface irrigation (23.77 t/ha). The surface irrigation method resulted in considerably low mean crop yield, which may be due to poor soil distribution and nonuniform distribution of soil moisture.

Irrigation levels and irrigation methods had marked effect on IPE of cauliflower. The IPE for different irrigation levels ranged from 8.03 to 29.39 kg/m³. The IPE was decreased significantly with the increase in irrigation levels. Irrigation at 25% of pan evaporation replenishment resulted in higher mean IPE (29.39 kg/m³), because the reduction in seasonal water depth was higher than the reduction in crop yield. A further increase in irrigation level from 25% to 175% of pan evaporation replenishment reduced the IPE significantly, because the increase in crop yield was less than increase in seasonal water depth. Irrigation at 175% of pan evaporation replenishment resulted in significantly minimum IPE (8.03 kg/m³), because of increase in the seasonal water application depth considerably and the decrease in crop yield. Irrigation methods significantly influenced the IPE of cauliflower. The IPE of cauliflower for drip and surface irrigation methods were 14.2 and 10.24 kg/m³, respectively.

**TABLE 3.4**   Effects of Irrigation Scheduling and Irrigation Methods on Marketable Yield, Yield Components, and Irrigation Production Efficiency of Cauliflower.

| | Mean yield (t/ha) | Mean irrigation production efficiency (kg/m3) |
|---|---|---|
| **Irrigation scheduling (pan evaporation replenishment, %)** | | |
| 25 | 17.05 | 29.39 |
| 75 | 26.43 | 15.18 |
| 125 | 37.66 | 12.98 |
| 175 | 32.61 | 8.03 |
| LCD (0.05) | 0.947 | 0.602 |
| **Irrigation methods** | | |
| Drip | 33.1 | 14.2 |
| Surface | 23.77 | 10.24 |
| LCD (0.05) | 0.670 | 0.426 |
| **Interaction (irrigation scheduling × methods)** | | |
| LCD (0.05) | 1.34 | 0.851 |

Drip irrigation method resulted in significantly higher IPE, due to higher crop yield. Surface irrigation resulted in minimum IPE, due to considerably low mean crop yield. The overall results presented in Table 3.4 revealed that both irrigations levels and methods influenced the mean crop yield and mean IPE of cauliflower. The highest mean crop was recorded when irrigation during the crop growing season was applied at 125% of Ep, whereas mean IPE was higher with irrigation at 25% of Ep.

### 3.4.2.2   WATER APPLICATION DEPTH AND MARKETABLE YIELD OF CAULIFLOWER

The relationship between seasonal water application depth and marketable yield of cauliflower for different irrigation methods are represented in Figure 3.10. The seasonal water application depth varied from 58 mm to 406 mm, whereas crop yield ranged from 21.20 to 41.60 t/ha under drip irrigation and 12.90 to 33.72 t/ha under surface irrigation, respectively. The seasonal water application depth versus crop yield for drip system ($R^2$ = 0.970) and surface irrigation ($R^2$ = 0.876) exhibited strong quadratic relationships. Cauliflower maximum marketable yield was obtained with seasonal water depth of 290 mm under drip and 300 mm under surface irrigation methods, respectively, and thereafter it tended to decline.

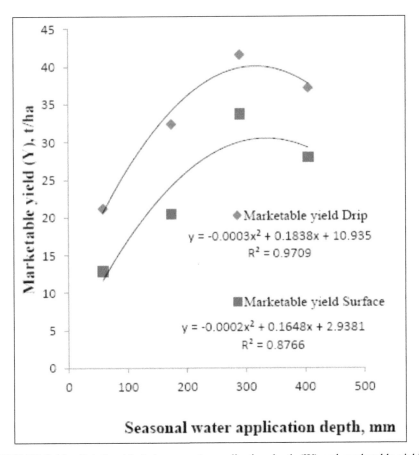

**FIGURE 3.10**  Relationship between water application depth (W) and marketable yield (Y) of cauliflower.

The relationships between marketable yield and pan evaporation replenishment are presented in Figure 3.11. The pan evaporation replenishment and crop yield of cauliflower for drip system ($R^2$ = 0.969) and surface irrigation system ($R^2$ = 0.876) exhibited strong quadratic relationships. The marketable yield was increased with an increase in pan evaporation replenishment up to 125 under drip to 135% under surface irrigation methods; and there after crop yield tended to decline.

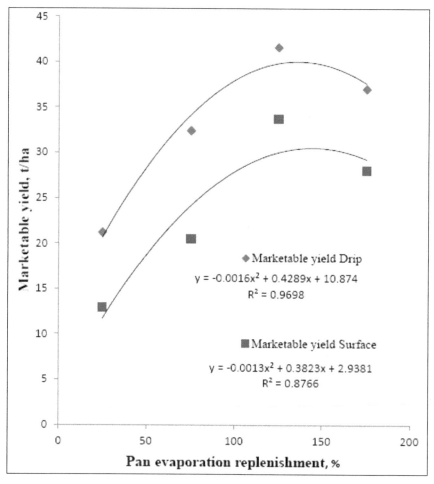

**FIGURE 3.11** Relationship between pan evaporation replenishment and marketable yield (Y) of cauliflower.

### 3.4.2.3 ECONOMIC ANALYSIS: TOTAL COST OF PRODUCTION, GROSS RETURN, NET RETURN, AND BENEFIT COST RATIO FOR CAULIFLOWER CROP

The total cost of production, gross return, net return and BCR of cauliflower in relation to irrigation methods and irrigation schedules are presented in Table 3.5. The price of cauliflower was assumed as 10 Rs./kg. The procedure described in Appendices A2–A7 were used to estimate economic returns of cauliflower production under two irrigation methods.

**TABLE 3.5** Economic Returns of Cauliflower under Different Irrigation Scheduling and Irrigation Methods.

| Treatment (pan evaporation replenishment) (%) | Total cost of production (Rs./ha) | | Gross return (Rs./ha) | | Net return (Rs./ha) | | Benefit cost ratio (BCR) | |
|---|---|---|---|---|---|---|---|---|
| | Drip | Surface | Drip | Surface | Drip | Surface | Drip | Surface |
| 25 | 86,341 | 63,706 | 212,000 | 129,000 | 125,658 | 65,294 | 2.45 | 2.02 |
| 75 | 87,489 | 64,855 | 324,000 | 204,333 | 237,843 | 139,479 | 3.68 | 3.11 |
| 125 | 88,637 | 66,003 | 416,000 | 337,000 | 323,362 | 270,996 | 4.64 | 5.10 |
| 175 | 89,786 | 67,156 | 372,000 | 281,333 | 283,546 | 214,182 | 4.13 | 4.12 |

The total cost of production was increased slightly with increase in irrigation levels due to insignificant increase in pumping cost induced by variation in seasonal water application depth. The total cost varied from 63,706 to 67,156 Rs./ha for surface and 86,341 to 89,786 Rs./ha for drip irrigation methods, depending on pan evaporation replenishment. The cost of production in drip irrigation is considerably higher in surface irrigation mainly due to variation in irrigation system cost. The gross return for surface and drip conditions ranged from 129,000 to 337,000 Rs./ha and 212,000 to 416,000 Rs./ha, respectively, under different irrigation scheduling treatments. The gross return was increased sharply with increase from 25% to 125% pan evaporation replenishment due to significant increase in marketable yield. Irrigation at 175% of pan evaporation replenishment reduced the gross return due to significant reduction in marketable yield. The net return for surface and drip irrigation method at 25% of pan evaporation replenishment gave the least values. Then the irrigation from 75% to 125% pan evapo-ration replenishment increased the yield sharply. A further increase in irrigation amount to 175% of pan evaporation replenishment reduced the net returns due to considerable reduction in marketable yield. The maximum net returns of 270,966 and 323,362 Rs./ha were obtained at 125% of pan evaporation replenishment for surface and drip irrigation, respectively (Table 3.5).

The BCR for surface and drip condition ranged from 2.02 to 5.10 and 2.45 to 4.64, respectively, under different irrigation levels (Table 3.5). The BCR was increased considerably with increasing pan evaporation replenishment from 25% to 125%. Irrigation at 175% of pan evaporation replenishment decreased the BCR, because it increased the total cost of production but decreased the marketable yield.

The overall results clearly revealed that the irrigation at 125% pan evaporation replenishment gave the higher net return, gross return and BCR. The results further revealed that drip irrigation gave higher economic return than surface irrigation condition.

## 3.4.2.4  IRRIGATION DEPTH AND ECONOMIC RETURNS

The relationships between seasonal water application depths and gross return of cauliflower under drip and surface irrigation systems are shown in Figure 3.12.

The seasonal water application ranged from 58 to 406 mm (Table 3.6); and gross returns for drip irrigation ($R^2 = 0.970$) and surface irrigation ($R^2 = 0.877$) exhibited strong quadratic relationships. The gross returns of cauliflower were increased with the increase in seasonal water application from 290 to 350 mm for drip and surface irrigation methods, respectively, and thereafter gross return tended to decline (Fig. 3.12).

The relationship between pan evaporation replenishment and gross return of cauliflower under drip and surface irrigation are shown in Figure 3.13. The pan evaporation replenishment versus gross return show the strong quadratic relationships for drip ($R^2 = 0.991$) and surface ($R^2 = 0.877$). The gross return was increased with increasing pan evapo-ration replenishment up to 150% in case of drip and surface irrigation methods, respectively, and thereafter it tended to decline (Fig. 3.13).

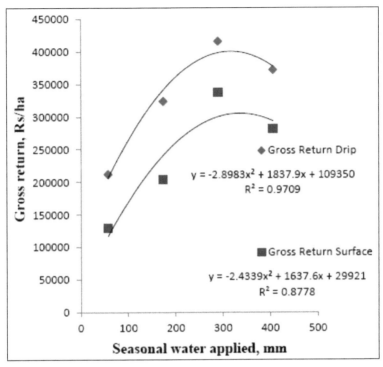

**FIGURE 3.12**   Relationship between water application depth and gross return (GR) of cauliflower.

**TABLE 3.6**   Effects of Irrigation Scheduling Treatments and Seasonal Water Application Depths on Marketable Yield and Irrigation Production Efficiency of Cauliflower.

| Treatment | Seasonal water applied, mm | Marketable yield, t/ha | Irrigation production efficiency, kg/m3 |
|---|---|---|---|
|  |  | **Mean** |  |
| 0.25 EPD | 58 | 21.20 | 36.54 |
| 0.75 EPD | 174 | 32.40 | 18.61 |
| 1.25 EPD | 290 | 41.60 | 14.34 |
| 1.75 EPD | 406 | 37.2 | 9.16 |
| 0.25 EPS | 58 | 12.90 | 22.2 |
| 0.75 EPS | 174 | 20.46 | 11.73 |
| 1.25 EPS | 290 | 33.72 | 36.54 |
| 1.75 EPS | 406 | 28.03 | 18.61 |

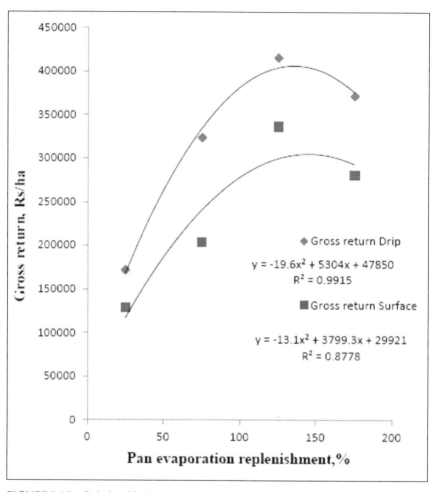

**FIGURE 3.13** Relationship between pan evaporation replenishment (%) and gross return (GR) of cauliflower.

The relationships between seasonal water application and net returns of cauliflower under drip and surface irrigation are shown in Figure 3.14. Seasonal water application and net returns for drip irrigation ($R^2 = 0.977$) and surface irrigation ($R^2 = 0.842$) exhibited strong quadratic relationships. The net returns of cauliflower were increased with increase in seasonal water application up to 290 mm for drip and 300 mm for surface irrigation and thereafter it tended to decline (Fig. 3.14).

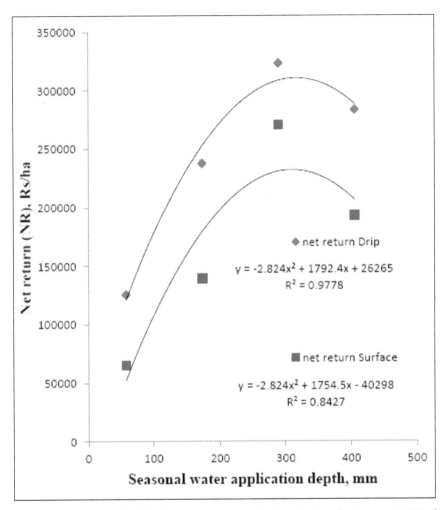

**FIGURE 3.14**   Relationships between water application depth and net return (NR) of cauliflower.

The relationships between pan evaporation replenishment and net return of cauliflower for drip and surface irrigation are presented in Figure 3.15. The pan evaporation replenishment and net return show strong quadratic relationships for drip ($R^2 = 0.994$) and surface ($R^2 = 0.842$) irrigation. The net return was increased with increasing pan evaporation replenishment up to 125% in both irrigation methods and thereafter at 175% it tended to decline.

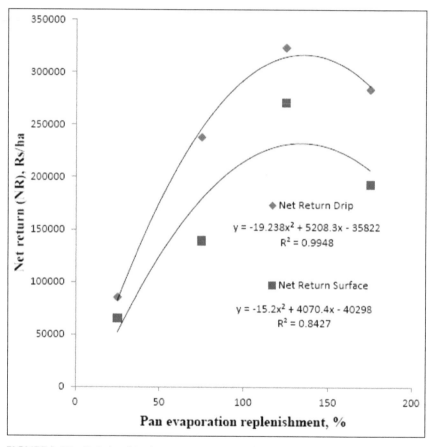

**FIGURE 3.15**   Relationships between pan evaporation replenishment (%) and net return (NR) of cauliflower.

The relationships between seasonal water application and BCR of cauliflower under drip and surface irrigation are shown in Figure 3.16. The seasonal water application and BCR of cauliflower for drip irrigation ($R^2 = 0.972$) and surface irrigation ($R^2 = 0.857$) exhibited strong quadratic

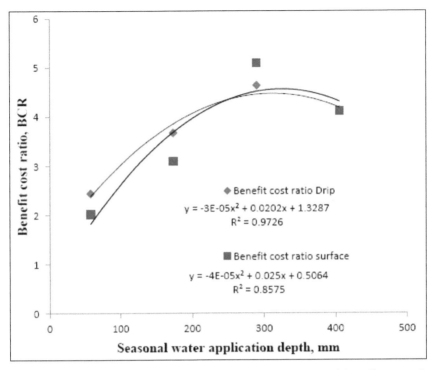

**FIGURE 3.16** Relationships between water application depth and benefit cost ratio (BCR) of cauliflower.

relationship. The BCR of cauliflower was increased with an increase in seasonal water application up to 290 mm for drip and 300 mm for surface irrigation and thereafter it tended to decline (Fig. 3.16).

The relationships between pan evaporation replenishment and BCR of cauliflower under drip and surface irrigation are shown in Figure 3.17. The pan evaporation replenishment and BCR show strong quadratic relationship for drip irrigation ($R^2 = 0.972$) and surface irrigation ($R^2 = 0.857$). The BCR was increased with an increase in pan evaporation replenishment up to 125% in case of drip and surface irrigation and thereafter it tended to decline.

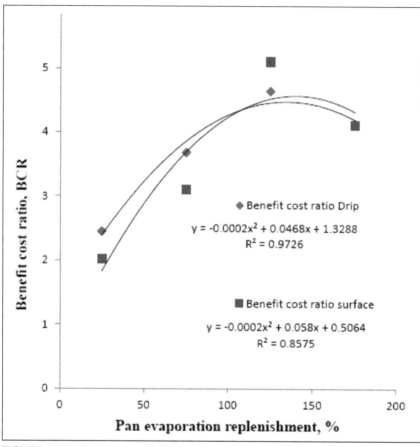

**FIGURE 3.17** Relationships between pan evaporation replenishment (%) and benefit cost ratio (BCR) of cauliflower.

### 3.4.2.5 CONCLUSIONS

1. The experimental results show that irrigation at 125% of pan evaporation replenishment gave a significantly maximum yield of cauliflower due to higher crop weight, whereas irrigation at 25% of pan evaporation replenishment resulted in highest IPE.
2. Drip irrigation method resulted in higher crop yield and IPE than the surface irrigation. Although drip irrigation method is costly,

yet the results clearly indicate higher profits due to high production and water efficiency of cauliflower.

3. Finally, the overall results clearly suggest that in order to obtain higher crop yield, IPE and net return of cauliflower during the winter growing season (December to March), the crop should be irrigated at 125% of pan evaporation replenishment under drip irrigation.

### 3.4.3   TOMATO

This section of research study on onion was conducted by Suchit Ekka, Niman Bodra, and Shalini Sharma, graduate students under the supervision of Professor S. K. Srivastava.

#### 3.4.3.1   YIELD AND IRRIGATION PRODUCTION EFFICIENCY OF TOMATO

The effects of irrigation scheduling on marketable yield and IPE of tomato are shown in Table 3.7. The marketable yield of tomato ranged from 25.7 to 52.54 t/ha among different treatments. The marketable yield of tomato was increased significantly with an increase in irrigation level up to 125% of pan evaporation replenishment with a maximum marketable yield of 52.54 t/ha. A further increase in irrigation level up to 175% of pan evaporation replenishment reduced the marketable yield to 49.98 t/ha significantly due to reduction in mean fruit weight.

The irrigation levels significantly influenced the IPE of tomato (Table 3.7). IPE was decreased significantly with an increase in irrigation level, because of increase in seasonal water application depth. The irrigation at 25% of pan evaporation replenishment resulted in significant maximum IPE of 22.35 kg/m$^3$ because of reduction in seasonal water application depth. Irrigation at 175% of pan evaporation replenishment resulted in significantly minimum IPE of 6.21 kg/m$^3$ because it increased the seasonal water application depth but at the same time decreased the marketable yield.

**TABLE 3.7**   Effects of Irrigation Scheduling and Irrigation Methods on Marketable Yield, Yield Components, and Irrigation Production Efficiency of Tomato.

| Treatment | Mean yield (t/ha) | Mean irrigation production efficiency (kg/m³) |
|---|---|---|
| **Irrigation scheduling (pan evaporation replenishment, %)** | | |
| 25 | 25.7 | 22.35 |
| 75 | 44.55 | 12.95 |
| 125 | 52.54 | 9.14 |
| 175 | 49.98 | 6.21 |
| CD (0.05) | 0.209 | 0.690 |
| **Irrigation methods** | | |
| Drip | 49.44 | 14.83 |
| Surface | 36.95 | 10.48 |
| CD (0.05) | 0.148 | 0.488 |
| **Interaction (irrigation scheduling × methods)** | | |
| CD (0.05) | 0.296 | 0.975 |

## 3.4.3.2   ECONOMIC ANALYSIS: TOTAL COST OF PRODUCTION, GROSS RETURN, NET RETURN, AND BENEFIT COST RATIO FOR TOMATO CROP

The total cost of production, gross return, net return and BCR of tomato versus irrigation methods and scheduling are presented in Table 3.8. The procedure described in Appendices A2–A7 was used to estimate economic returns of tomato production under two irrigation methods. The total cost of production was increased slightly with an increase in irrigation level due to increase in pumping cost induced by variation in seasonal water application depth. The total cost of production varied from 101,711 to 112,061 t/ha for drip and 71,507 to 81,857 t/ha for surface irrigation, respectively. The total cost of production under drip irrigation was significantly higher as compared with surface irrigation mainly due to high cost of drip irrigation system. The gross return under different irrigation levels ranged from 406,207 to 730,340 Rs./ha for drip and 262,123 to 569,140 Rs./ha for surface irrigation, respectively. The increase in gross return at 125% of pan evaporation replenishment was due to considerable higher marketable yield. A further increase in irrigation level up to 175% of pan evaporation replenishment decreased the gross return considerably due to reduction in marketable yield. The gross return of surface irrigation was considerably lower than the drip irrigation due to lower marketable yield induced by poor water distribution. The net return was increased considerably with an increase in irrigation level. The maximum net return for drip (655,788 Rs./ha) and surface (523,276 Rs./ha) irrigation methods were obtained when irrigation during crop growing season was applied at 125% of pan evaporation replenishment. A further increase in irrigation up to 175% of pan evaporation replenishment reduced the net return considerably due to reduction in gross return. In spite of lower system cost, the surface irrigation method gave considerably low net return as compared with drip irrigation systems mainly due to lower gross return.

**TABLE 3.8** Economic Return of Tomato under Different Irrigation Scheduling and Irrigation Methods.

| Treatment (pan evaporation replenishment) (%) | Total cost of production (Rs./ha) | | Gross return (Rs./ha) | | Net return (Rs./ha) | | Benefit cost ratio (BCR) | |
|---|---|---|---|---|---|---|---|---|
| | Drip | Surface | Drip | Surface | Drip | Surface | Drip | Surface |
| 25 | 101,711 | 71,507 | 406,207 | 262,123 | 304,495 | 190,616 | 3.99 | 5.60 |
| 75 | 105,161 | 74,957 | 669,933 | 488,540 | 564,772 | 413,583 | 6.37 | 7.52 |
| 125 | 108,611 | 78,407 | 764,400 | 601,683 | 655,789 | 523,276 | 7.04 | 8.67 |
| 175 | 112,061 | 81,857 | 730,340 | 569,140 | 618,279 | 487,283 | 6.52 | 7.95 |

The BCR ranged from 3.99 to 7.04 for drip and 5.60 to 8.67 for surface irrigation, respectively. The BCR was increased with an increase in irrigation levels up to 125% of pan evaporation replenishment due to significant increase in gross return. A further increase in irrigation level up to 175% of pan evaporation replenishment reduced the total cost of production. The drip irrigation system resulted in higher BCR followed by surface irrigation system (Table 3.8).

The overall results revealed that irrigation at 125% of pan evaporation replenishment resulted in higher gross return, net return, and BCR. The results further revealed that drip irrigation system resulted in higher gross return, net return, and BCR.

### 3.4.3.3   WATER APPLICATION DEPTH AND TOMATO YIELD

The relationship between seasonal water application depth and marketable yield of tomato for drip and surface irrigation methods are presented in Figure 3.18 and Table 3.9. The statistical analysis for the data is shown in the ANOVA Table 3.10.

In spite of some variation, the seasonal water application depth and marketable yield of tomato for drip ($R^2 = 0.999$) and surface ($R^2 = 0.9993$) irrigation methods exhibited strong quadratic relationships. The marketable yield of tomato was increased with increase in seasonal water application depth up to 575 mm for drip and surface irrigation methods, respectively, and thereafter, yield tended to decline (Fig. 3.18).

The relationship between pan evaporation replenishment (irrigation scheduling) and marketable yield of tomato under drip and surface irrigation methods are presented in Figure 3.19. The pan evaporation replenishment and marketable yield of tomato for drip ($R^2 = 0.999$) and surface ($R^2 = 0.9993$) irrigation methods exhibited strong quadratic relationships. The tomato yield was the maximum at 125% of pan evaporation replenishment for drip and surface irrigation methods, respectively and thereafter the yield tended to decline (Fig. 3.19).

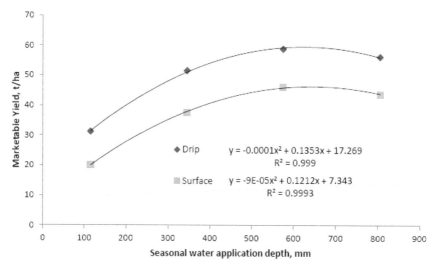

**FIGURE 3.18**   Relationships between seasonal water application depth and marketable yield of tomato for drip (◆) and surface (■) irrigation methods.

**TABLE 3.9** Effects of Irrigation Scheduling and Irrigation Methods on Seasonal Water Application Depth and Marketable Yield of Tomato.

| Treatment | Seasonal water applied, mm | Marketable yield, t/ha | Irrigation production efficiency, kg/m3 |
|---|---|---|---|
| | | | Mean |
| 0.25 EPD | 115 | 27.17 | 31.24 |
| 0.75 EPD | 345 | 14.94 | 51.53 |
| 1.25 EPD | 575 | 10.23 | 58.80 |
| 1.75 EPD | 805 | 6.98 | 56.18 |
| 0.25 EPS | 115 | 17.53 | 20.16 |
| 0.75 EPS | 345 | 10.89 | 37.58 |
| 1.25 EPS | 575 | 8.05 | 46.28 |
| 1.75 EPS | 805 | 5.44 | 43.78 |

**TABLE 3.10** Analysis of Variance: Effects of Irrigation Schedule and Irrigation Methods on Seasonal Water Application Depth and Irrigation Production Efficiency of Tomato.

| Source | d.f. | S.S. | M.S.S. | F. Cal. | F.Tab. 5% | Result | S. Em. (±) | C.D. at 5% |
|---|---|---|---|---|---|---|---|---|
| Replication | 2 | 2.079 | 1.039 | 3.06 | 3.74 | NS | – | – |
| Due to irrigation schedule (I) | 3 | 724.46 | 241.49 | 710.1 | 3.34 | S | 0.337 | 0.690 |
| Due to irrigation Methods (M) | 1 | 2057 | 2057 | 6051 | 4.60 | S | 0.238 | 0.488 |
| Interaction (I × M) | 3 | 71.98 | 23.99 | 70.55 | 3.34 | S | 0.476 | 0.975 |
| Error | 14 | 4.76 | 0.34 | – | – | – | – | – |
| Total | 23 | 2861 | – | – | – | – | – | – |

The quadratic relationships among yield versus water application depth probably resulted from nutrients leaching through deep percolation and poor aeration. These results are in agreement with Imtiyaz et al.[11–15] for cabbage, broccoli, tomato, cauliflower, onion, spinach and carrot under sprinkler and drip irrigation methods. Kumar et al.[18] and other researchers have also reported quadratic relationships between yield and seasonal water application depth for selected vegetable crops under wide variety of irrigation systems and regimes, soil and climatic conditions.

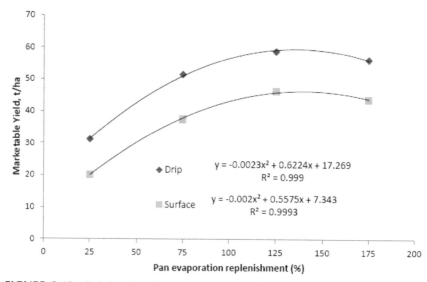

**FIGURE 3.19** Relationships between pan evaporation replenishment and marketable yield of tomato for drip (◆) and surface (■) irrigation methods.

### 3.4.3.4   WATER APPLICATION AND ECONOMIC RETURN OF TOMATO

The relationship between seasonal water application depth and gross return of tomato under drip and surface irrigation methods are presented in Figure 3.20. The seasonal water application depth and gross return of tomato under drip ($R^2 = 0.999$) and surface ($R^2 = 0.9993$) irrigation methods exhibited strong quadratic relationships. The gross return was increased with an increase in seasonal water application depth up to 575 mm for drip and surface irrigation methods, respectively and thereafter gross return tended to decline. The results revealed that higher seasonal water application beyond above mentioned values did not increase the gross return (Fig. 3.20).

The relationships between pan evaporation replenishment and gross return of tomato for drip and surface irrigation methods are presented in Figure 3.21. The pan evaporation replenishment and gross return of tomato under drip ($R^2 = 0.999$) and surface ($R = 0.9993$) irrigation methods exhibited strong quadratic relationships. The tomato attained the maximum gross return at 125% of pan evaporation replenishment for drip and surface

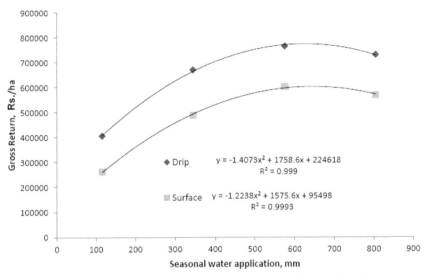

**FIGURE 3.20**   Relationships between seasonal water application depth and gross return of tomato for drip ($\blacklozenge$) and surface ($\blacksquare$) irrigation methods.

irrigation methods, respectively, and thereafter the gross return tended to decline (Fig. 3.21).

The relationships between seasonal water application depth and net return of tomato under drip and surface irrigation methods are illustrated in Figure 3.22. The seasonal water application depth and net return of tomato under drip ($R^2 = 0.9989$) and surface ($R^2 = 0.9992$) irrigation methods exhibited strong quadratic relationships. The tomato attained the maximum net return at 575 mm of seasonal water application for drip and surface irrigation methods, respectively and thereafter the net return tended to decline (Fig. 3.22).

The relationships between pan evaporation replenishment and net return of tomato under drip and surface irrigation methods are illustrated in Figure 3.23. The pan evaporation replenishment and net return of tomato under drip ($R^2 = 0.9989$) and surface ($R^2 = 0.9992$) irrigation methods exhibited strong quadratic relationships. The tomato attained the maximum net return at 125% of pan evaporation replenishment for drip and surface irrigation methods, respectively and thereafter the net return tended to decline (Fig. 3.23).

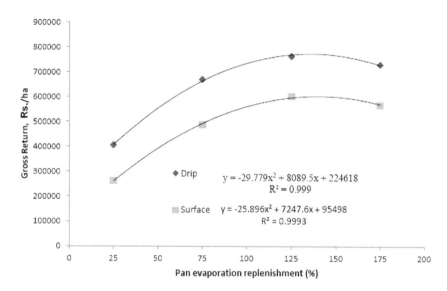

**FIGURE 3.21**   Relationships between pan evaporation replenishment and gross return of tomato for drip (◆) and surface (■) irrigation methods.

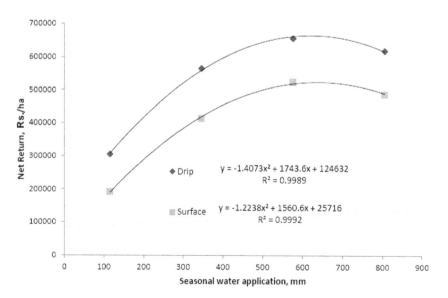

**FIGURE 3.22**   Relationship between seasonal water application depth and net return of tomato for drip (◆) and surface (■) irrigation methods.

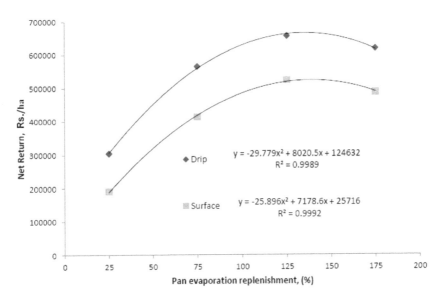

**FIGURE 3.23**   Relationships between pan evaporation replenishment and net return of tomato for drip (◆) and surface (■) irrigation methods.

The relationships between seasonal water application depth and BCR of tomato under drip and surface irrigation methods are illustrated in Figure 3.24. The seasonal water application depth and BCR of tomato under drip ($R^2$ = 0.9976) and surface ($R^2$ = 0.9883) irrigation methods exhibited strong quadratic relationships. The tomato attained the maximum net BCR at 575 mm for drip and surface irrigation methods, respectively, and thereafter the BCR tended to decline (Fig. 3.24).

The relationships between pan evaporation replenishment and BCR of tomato under drip and surface irrigation methods are illustrated in Figure 3.25. The pan evaporation replenishment and BCR of tomato under drip ($R^2$ = 0.9976) and surface ($R^2$ = 0.9883) irrigation methods exhibited strong quadratic relationships. The tomato attained the maximum BCR at 125% of pan evaporation replenishment for drip and surface irrigation methods respectively and thereafter, the BCR tended to decline (Fig. 3.25).

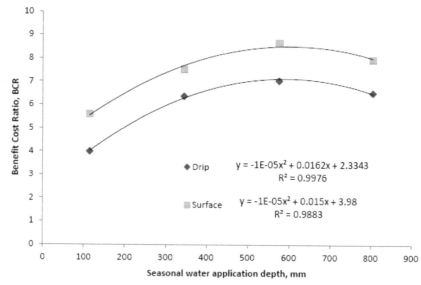

**FIGURE 3.24** Relationships between seasonal water application depth and benefit cost ratio of tomato for drip (◆) and surface (■) irrigation methods.

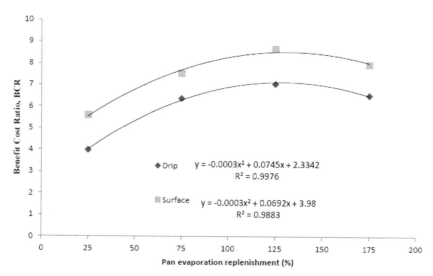

**FIGURE 3.25** Relationships between pan evaporation replenishment and benefit cost ratio of tomato for drip (◆) and surface (■) irrigation methods.

### 3.4.3.5  CONCLUSIONS

1. Irrigation at 125% of pan evaporation replenishment resulted in higher gross return, net return and BCR of tomato crop.
2. Irrigation at 125% of pan evaporation replenishment resulted in significantly higher marketable yield of tomato whereas IPE was higher with irrigation at 25% of pan evaporation replenishment for tomato. A further increase in irrigation level up to 175% of pan evaporation replenishment reduced both marketable yield and IPE.
3. The seasonal water application depth/irrigation levels of marketable yield, gross return, net return and BCR exhibited strong quadratic relationships. This result can be used for optimizing marketable yield and economic return of tomato under limited water resources.
4. Finally, the overall results revealed that in order to obtain higher marketable yield and economic return of tomato, the crop should be irrigated at 125% of pan evaporation replenishment under the prevailing climatic conditions of the region in this chapter. Furthermore, in spite of high initial investment of the drip irrigation system, tomato production is highly profitable.

## 3.5  SUMMARY

The present study was conducted with the following objectives to improve marketable yield, water-use efficiency and economic return of onion, cauliflower and tomato under drip, and surface irrigation methods:

1. To investigate the effects of irrigation scheduling on yield and IPE of onion, cauliflower and tomato.
2. To investigate the effects of irrigation scheduling on economic returns of onion, cauliflower and tomato.
3. To develop water-yield relationships of onion, cauliflower and tomato in order to optimize the yield under limiting water supply conditions.
4. To develop relationships between seasonal water application/irrigation scheduling and crop yield, gross return, net return, and BCR in order to maximize the profits: Onion, cauliflower and tomato.

## ACKNOWLEDGMENT

At Vaugh School of Agricultural Engineering and Technology, SHIATS, Allahabad, India, author expresses his gratitude to his M.Tech. students, namely: Abhay Bara and Suman Tigga for research on onion crop; PratimaHoro, ShailendraTirkey, and Geoffrey Bai Passah for research on cauliflower crop; and Suchit Ekka, NimanBodra, and Shalini Sharma for research on tomato.

## KEYWORDS

- **benefit cost ratio**
- **capital recovery factor**
- **crop water production function**
- **drip irrigation**
- **net return**

## REFERENCES

1. Adamese, F. J. Irrigation Methods and Water Quality Effects on Onion Yield and Grade. *Agron. J.* **1989,** *81*(4), 589–553.
2. Arkley, A. J. Relationship between Plant Growth and Transpiration. *Hilgaradia.* **1963,** *34,* 559–584.
3. Bastiaanssen, W. G. M.; Singh, R.; Kumar, S.; Agrawal, M. C. Control of Soil Degradation by Modified and Drainage Techniques. In *Impact of Modern Agriculture on Environment;* Behl, R. K., Arora, S. K., Tauro, P., Eds.; CCS HAU: Hisar, 1994; pp 83–94.
4. Bernstein, L.; Francois, L. E. Comparison of Drip, Furrow and Sprinkler Irrigation. *Soil Sci.* **1993,** *115,* 73–86.
5. Bockhold, D. L.; Thompson, A. L.; Sudduth, K. A.; Henggeler, J. C. Irrigation Scheduling Based on Crop Canopy Temperature for Humid Environment. *Trans. ASABE.* **2011,** *54*(6), 2021–2028.
6. Bucks, D. A.; French, O. F. Quantity and Frequency of Trickle and Furrow Irrigation for Efficient Cabbage Production. *Agron. J.* **1974,** *66,* 53–57.
7. Cetin, O.; Yidirim, O.; Hygan, D.; Boyaci, H. Irrigation Scheduling of Drip Irrigation Tomatoes Using Class A Pan Evaporation. *Turk. J. Agric.* **2002,** *26,* 171–178.
8. Chaudhary, P. N.; Kumar, V. The Sensitivity of Growth and Yield of Dwarf Wheat to Water Stress at Three Growth Stage. *Irrig. Sci.* **1980,** *1,* 223–227.

9. Curmen, D.; Massie, L. R. Potato Irrigation Scheduling in Wisconsin. *Am. Potato. J.* **1984,** *61*(4), 235–241.

10. Dingre, S. K.; Pawar, D. D.; Kadam, K. G. Productivity of Water Use and Quality of Onion Seed Production under Different Scheduling through Drip. *Indian J Agron.* **2012,** *57*(2), 186–190.

11. Imtiyaz, M.; Mgadla, N. P.; Manase, S. K. Response of Green Mealies to Water Levels under Sprinkler and Drip Irrigation. In *Proceedings of International Agricultural Engineering Conference;* Asian Institute of Technology: Bangkok, 2000; pp 343–350.

12. Imtiyaz, M.; Mgadla, N. P.; Chepete, B. *Yield and Water Expense Efficiency of Onion, Tomato and Green Pepper as Influenced by Irrigation Scheduling, Irrigation Research Report 3;* Department of Agricultural Research: Botswana, 1995; p 25.

13. Imtiyaz, M.; Mgadla, N. P.; Chepete, B.; Manase, S. K. Marketable Yield. In *Economics of Water Resources Planning;* James, L. D., Lee, R. R., Eds.; McGraw Hill Book Co.: New Delhi, 2000; p 20.

14. Imtiyaz, M.; Mgadla, N. P.; Chepete, B.; Manase, S. K. Response of Six Vegetable Crops to Irrigation Scheduling. *Agric. Water Manag.* **2000,** *45,* 331–342.

15. Imtiyaz, M.; Mgadla, N. P.; Manase, S. K.; Chendo, K.; Mothobi, E. O. Yield and Economic Return of Vegetable Crops under Variable Irrigation. *Irrig. Sci.* **2000,** *19,* 87–93.

16. Jensen, M. E.; Robb, D. C. N.; Franzoy, C. E. Scheduling Irrigation Using Climate Crop Soil Data. *J. Irrig. Drain Div.* **1970,** *96,* 25–38.

17. Khade, K. K. *Highlights of Research on Drip Irrigation*; Mahatma Phule Agricultural University: Rahuri, Ahmednagar, Pub. No. 55, 1987; pp 20–21.

18. Kumar. A.; Singh, R. Response of Onion (*Allium cepa* L.) to Different Levels of Irrigation Water. *Agric. Water Manag.* **2007,** *89*(1–2), 161–166.

19. Mgadla, N. P.; Imtiyaz, M.; Chepete, B. *Wheat Production as Influenced by Limited Irrigation, Irrigation Research Paper 2;* Department of Agricultural Research: Botswana, 1995; p 22.

20. Mishra, R. D.; Plant, P. C. Criteria for Scheduling Irrigation of Wheat. *Exp. Agric.* **1981,** *17*(2), 22.

21. Smidt, D. *Proceedings of Indo-German Conference on Impact of Modern Agriculture on Environment.* CCS Haryana Agricultural University: Hisar, 1993; pp 83–93.

22. Sammis, T. Comparison of Sprinkler, Trickle, Sub-surface and Furrows Irrigation Methods for Row Crops. *Agron. J.* **1979,** *72*(5), 701–704.

23. Sammis, T.; Sharma, P.; Shukla, M. K; Wang, J.; Miller, D. A Water Balance Drip Irrigation Scheduling Model. *Agric. Water Manag.,* **2012,** *113*(10), 30–37.

24. Singh, P. N.; Mohan, S. C. Water Use and Yield Response of Sugarcane under Different Irrigation Scheduling and Nitrogen Levels in Sub-tropical Region. *Agric. Water Manag.* **1994,** *26,* 253–264.

25. Singh, S.; Ram, D.; Sharma, S.; Singh, D. V. Water Requirements and Productivity of Palmarosa on Sandy Loam Soil under Sub-tropical Climate. *Agric. Water Manag.* **1997,** *35,* 1–10.

26. Sivanappan, R. K.; Padmakumari, O.; Kumar, V. *Drip Irrigation;* Keerthi Publishing House: Coimbatore, 1997.

27. Srivastava, P. K.; Parikh, M. M.; Swami, N. G.; Raman, S. Effect of Drip Irrigation and Mulching on Tomato Yield. *Agric. Water Manag.* **1994,** *25,* 179–184.
28. Yield, Water Use Efficiency and Economic Return of Cabbage, Carrot and Onion as Influenced by Irrigation Scheduling. In *Proceedings of International Agricultural Engineering Conference;* Asian Institute of Technology: Bangkok; pp 321–328.
29. Zang, H.; Oweis, T. Water-yield Relations and Optimal Irrigation Scheduling of Wheat in the Mediterranean Region. *Agric. Water Manag.* **1999,** *38,* 195–211.

## APPENDICES A

**APPENDIX A1** Effects of Irrigation Scheduling Levels and Irrigation Methods on Seasonal Water Application and Marketable Yield of Onion.

| Treatment | Seasonal water application, mm | Marketable yield of onion, kg/ha | | | |
|---|---|---|---|---|---|
| | | 1 | 2 | 3 | Mean |
| 0.25 EP D | 97 | 17.3 | 17.5 | 16.8 | 17.20 |
| 0.75 EP D | 291 | 25.5 | 26.8 | 26 | 26.1 |
| 1.25 EP D | 485 | 33.1 | 33.1 | 29.2 | 33.4 |
| 1.75 EP D | 679 | 29.3 | 31.2 | 29.2 | 29.9 |
| 0.25 EP S | 97 | 13.1 | 13.1 | 14 | 13.4 |
| 0.75 EP S | 291 | 16.8 | 17.2 | 17.6 | 17.2 |
| 1.25 EP S | 485 | 21.1 | 22.1 | 22.1 | 21.7 |
| 1.75 EP S | 679 | 18.8 | 19.8 | 19.3 | 19.3 |

**APPENDIX A2** Effects of Irrigation Scheduling and Irrigation Methods on Seasonal Water Application and Irrigation Production Efficiency of Onion.

| Treatment | Seasonal water application, mm | Irrigation production efficiency of onion, kg/m3 | | | |
|---|---|---|---|---|---|
| | | 1 | 2 | 3 | Mean |
| 0.25 EP D | 970 | 17.36 | 17.35 | 18.48 | 17.73 |
| 0.75 EP D | 2910 | 9.22 | 8.84 | 8.82 | 8.96 |
| 1.25 EP D | 4850 | 7.81 | 6.61 | 6.22 | 6.88 |
| 1.75 EP D | 6790 | 5.1 | 4.0 | 4.1 | 4.4 |
| 0.25 EP S | 970 | 14.21 | 13.41 | 13.81 | 13.81 |
| 0.75 EP S | 2910 | 17.36 | 17.35 | 18.82 | 17.73 |
| 1.25 EP S | 4850 | 4.1 | 5.2 | 4.11 | 4.47 |
| 1.75 EP S | 6790 | 2.31 | 2.4 | 3.81 | 2.84 |

**APPENDIX A3**   Economic Analysis

## 1. Fixed cost

### (a) Water development

1) Tube well
2) Pump and motor
3) Pump and house
4) Water storage tanks
5) Main conveyance system
6) Other accessories (fitting etc.)

**Total cost = Rs. 421,500**

Useful life of tube well, pump and motor, pump house, and other accessories = 25 years

**Interest @12.5% per annum**

$$CRF = i(1+i)^n/(1+i)^n - 1 = 0.125(1+0.125)^{25}/(1+0.125)^{25} - 1 = 0$$

**Fixed cost Rs./year = CRF × Total cost** = $0.13194444 × 421,500$
= Rs. 55,614.58 (for 8 ha)

**Fixed cost/ha/year = 55,614.58/8 = 6951.82**

**Fixed cost/ha/yr/season = 6951.82/2 = 3475.91**

**Fixed cost = $R_s$. 3475.91----------------a**

### (b) Weeding and intercultural equipment

- Useful life = 7 years
- Sprayer = 6
- Hoes = 6
- Rake = 6
- Hand spade (Khurpi) = 7

Total cost = Rs. 8450

**Interest @12.5% per annum**

$$CRF = i(1+i)^n/(1+i)^n - 1 = 0.125(1+0.125)^7/(1+0.125)^7 - 1 = 0.21875$$

Fixed cost Rs./year = CRF × Total cost = $0.21875 × 8450$ = Rs. 1848.43

**Fixed cost/ha/season = Rs. 924.215----------------------b**

### (c1) Cost of irrigation system: Drip irrigation

- PVC pipe (Φ 50 mm) = 180m@40.50 Rs./m =$180 × 40.50$ = Rs. 7290
- LDPE Pipe (12 mm) = 10000m@9.5 Rs./m = Rs. 95,000

- Drippers = 45000/ha@4.05/pes = Rs. 182,250
- Fertilizer unit = Rs. 5500
- Filter = Rs. 3500
- Water meter = Rs. 2500
  **Sub-total cost = Rs. 296,040**
- **Accessories = Rs. 12.5% of subtotal** = Rs. 37,005

Total cost = Rs. 333,045

Interest per annum @12.5% and useful life of system = 8 year

$\mathbf{CRF = i(1 + i)^n/(1 + i)^n - 1} = 0.125(1 + 0.125)^8/(1 + 0.125)^8 - 1 = -0.204381$

Fixed cost ($R_s$/ha/year) = CRF × Total cost = 0.204381 × 333,045 = 68,068.07

**Fixed cost (Rs./ha/season) = Rs. 34,034.035**

Total cost of drip = a + b + c = 3475.91 + 924.21 + 34,034.035 = Rs. 38,434.155

**Total fixed cost for drip irrigation system = Rs. 38,434.15**
------------c

### (c2) Cost of irrigation system: Surface irrigation

Tank and concrete channel 250 m long cost = $R_s$ 3000

Interest rate @12.5 % per annum and useful life of system = 5 year

$\mathbf{CRF = i(1 + i)^n/(1 + i)^n - 1} = 0.125(1 + 0.125)^5/(1 + 0.125)^5 - 1 = 0.280860$

Fixed cost $R_s$/ha/year = CRF × Total cost = 0.280860 × 3000 = 842.58

Fixed cost $R_s$/ha/season = $R_s$ 842.58/2 = $R_s$ 421.29

Total fixed cost $R_s$/ha/season = 3475.91 + 924.21 + 421.29 = Rs. 4821.41-----------d

Total fixed cost surface irrigation system $R_s$/ha/season = Rs. 4821.41/ha/season-------------c

Total fixed cost of drip irrigation system $R_s$/ha/season = Rs. 38,434.15/ha/season-----------d

## 2. Operating cost

Energy meter = 9 kwh

Price = Rs. 4.75/kwh

Cost of energy/h = 9 × 4.75 = Rs. 42.75

Tube well discharge = 30 $m^3$/ha

Pumping cost = 42.75/30 = Rs. 1.425/$m^3$ of water

**Seasonal Water Application Depth and Pumping Cost:**

| Treatment | Seasonal water | | Pumping cost |
|---|---|---|---|
| | mm | m3/ha | |
| $0.25E_p$ | 97 | 970 | 1377.4 |
| $0.75E_p$ | 291 | 2910 | 4132.2 |
| $1.25E_p$ | 485 | 4850 | 6887 |
| $1.75E_p$ | 679 | 6790 | 9641.8 |

- **SEED (ONION DARK RED NASHIK N-53) Cost of seed = Rs. 400**
- **Fertilizer**
  Nitrogen = 25 kg/ha
  Urea (46% N) = 54.3 kg/ha @Rs. 9/kg; Cost of urea = Rs. 488.7
  $P_2O_5$ = 60 kg/ha; SSP (16%@P2O5) = 375 kg/ha @Rs. 8/kg; Cost
  of SSP = Rs. 3000
  $K_2O$ =100 kg/ha; MOP 60%$K_2O$ = 166.67 kg/ha @Rs. 8/kg; Cost of
  MOP = Rs. 1333.36
  **Total fertilizer cost = Rs. 4822.06**

- **Land preparation**
  Total cost = Rs. 2450/ha
  Workers (planting, preparation of bed, irrigation fertilizer applied,
  weeding and harvesting)
  Total labor cost = Rs. 20,000/ha
  Chemical = Rs. 600/ha
  Land rent = Rs. 8000/ha/season
  Land rent = Rs. 4000/ha/season
  Repair and maintains @2.5 of fixed cost
  Drip = Rs. 960.85/ha/year
  Surface = Rs. 120.53/ha/year
**Operating cost = Rs. 40,314.81--------------------e**

**APPENDIX A4** Effects of Irrigation Scheduling and Irrigation Methods on Fixed, Operating, and Total Cost of Onion Production.

| Treatment | Fixed cost (Rs./ha) | Operating cost (Rs./ha) | Total cost of production (Rs./ha) |
|---|---|---|---|
| | | From Appendix A3 | |
| 0.25 EP D | 38,434.15 | 42,653.06 | 81,087.21 |
| 0.75 EP D | 38,434.15 | 45,407.86 | 83,842.01 |
| 1.25 EP D | 38,434.15 | 48,162.66 | 86,596.81 |
| 1.75 EP D | 38,434.15 | 50,917.46 | 89,351.61 |
| 0.25 EP S | 4821.41 | 41,812.74 | 46,634.15 |
| 0.75 EP S | 4821.41 | 44,567.54 | 49,388.95 |
| 1.25 EP S | 4821.41 | 47,322.34 | 52,143.75 |
| 1.75 EP S | 4821.41 | 50,077.14 | 54,898.55 |

**APPENDIX A5** Effects of Irrigation Scheduling and Irrigation Methods on Gross Return of Onion.

| Treatment | Seasonal water application, mm | Gross return (Rs./ha) | | | |
|---|---|---|---|---|---|
| | | 1 | 2 | 3 | Mean |
| 0.25 EP D | 970 | 311,400 | 315,000 | 302,400 | 309,600 |
| 0.75 EP D | 2910 | 459,000 | 482,400 | 468,000 | 469,800 |
| 1.25 EP D | 4850 | 561,600 | 595,800 | 612,000 | 601,200 |
| 1.75 EP D | 6790 | 527,400 | 561,600 | 525,600 | 538,200 |
| 0.25 EPS | 970 | 235,800 | 235,800 | 252,000 | 241,200 |
| 0.75 EP S | 2910 | 302,400 | 309,600 | 316,800 | 309,600 |
| 1.25 EP S | 4850 | 379,800 | 397,800 | 397,800 | 390,600 |
| 1.75 EP S | 6790 | 338,400 | 347,400 | 347,400 | 347,400 |

**APPENDIX A6** Effects of Irrigation Scheduling and Irrigation Methods on Net Return of Onion.

| Treatment | Seasonal water application, mm | Net return (Rs./ha) | | | |
|---|---|---|---|---|---|
| | | 1 | 2 | 3 | Mean |
| 0.25 EP D | 97 | 230,312.8 | 233,912.8 | 221,312.8 | 228,512.8 |
| 0.75 EP D | 291 | 375,158 | 398,558 | 344,158 | 385,958 |
| 1.25 EP D | 485 | 475,003.2 | 509,203.2 | 525,403.2 | 514,603.2 |
| 1.75 EP D | 679 | 438,048.3 | 472,248.4 | 436,248 | 448,848.4 |
| 0.25 EP S | 97 | 189,165.9 | 189,165.9 | 205,365.9 | 194,565.9 |
| 0.75 EP S | 291 | 253,011.1 | 260,211.1 | 267,411.1 | 260,211.1 |
| 1.25 EP S | 485 | 327,656 | 345,656.3 | 345,656.3 | 338,436.3 |
| 1.75 EP S | 679 | 283,501.5 | 292,501.5 | 292,501.5 | 29,201.5 |

**APPENDIX A7** Effects of Irrigation Scheduling and Irrigation Methods on Benefit Cost Ratio of Onion.

| Treatment | Seasonal water application, mm | Benefit cost ratio (BCR) | | | |
|---|---|---|---|---|---|
| | | 1 | 2 | 3 | Mean |
| 0.25 EP D | 97 | 3.84 | 3.88 | 3.72 | 3.81 |
| 0.75 EP D | 291 | 5.47 | 5.75 | 5.58 | 5.6 |
| 1.25 EP D | 485 | 6.88 | 6.88 | 7.06 | 6.94 |
| 1.75 EP D | 679 | 5.9 | 6.28 | 5.88 | 6.02 |
| 0.25 EP S | 97 | 5.05 | 5.05 | 5.4 | 5.1 |
| 0.75 EP S | 291 | 6.12 | 6.26 | 6.41 | 6.26 |
| 1.25 EP S | 485 | 7.28 | 7.62 | 7.62 | 7.49 |
| 1.75 EP S | 679 | 6.26 | 6.32 | 6.32 | 6.42 |

## APPENDICES B: ANOVA TABLE

**APPENDIX B1**  Effects of Irrigation Scheduling and Irrigation Methods on Seasonal Water Application and Marketable Yield of Onion.

| Source | d.f | S.S. | M.S.S. | F. Cal | F. Tab 5% | Result | S.Em (±) | C.D. at 5% |
|---|---|---|---|---|---|---|---|---|
| Replication | 2 | 2.2433 | 1.6217 | 1.80 | 3.74 | NS | – | – |
| Due to irrigation scheduling (I) | 3 | 416.67 | 138.89 | 154.33 | 3.34 | S | 0.547 | 1.121 |
| Due to irrigation method (M) | 1 | 447.75 | 447.75 | 498.16 | 4.60 | S | 0.387 | 0.793 |
| INT (I × M) | 3 | 43.35 | 14.45 | 16.08 | 3.34 | S | 0.774 | 1.585 |
| Error | 14 | 12.58 | 0.90 | – | – | – | – | – |
| Total | 23 | 923.593 | – | – | – | – | – | – |

**APPENDIX B2**  Effects of Irrigation Scheduling and Irrigation Methods on Seasonal Water Application and IPE of Onion.

| Source | d.f | S.S. | M.S.S. | F. Cal | F. Tab 5% | Result | S.Em (±) | C.D. at 5% |
|---|---|---|---|---|---|---|---|---|
| Replication | 2 | 0.6203 | 0.3101 | 0.68 | 3.74 | NS | – | – |
| Due to irrigation scheduling (I) | 3 | 0.37 | 0.12 | 0.27 | 3.34 | NS | 0.390 | 0.800 |
| Due to irrigation methods (M) | 1 | 622.12 | 622.12 | 1359.99 | 4.6 | S | 0.276 | 0.565 |
| Int (I × M) | 3 | 153.41 | 51.14 | 111.79 | 3.34 | S | 0.552 | 1.131 |
| Error | 14 | 6.40 | 0.46 | – | – | – | – | – |
| Total | 23 | 782.926 | – | – | – | – | – | – |

# CHAPTER 4

# TENSIOMETER-BASED IRRIGATION SCHEDULING: DRIP-IRRIGATED BELL PEPPER UNDER NATURALLY VENTILATED POLYHOUSE

ASHWANI KUMAR MADILE and P. K. SINGH*

*Department of Irrigation and Drainage Engineering, G. B. Pant University of Agriculture and Technology, Pantnagar 263145, Udham Singh Nagar, Uttarakhand, India*

*Corresponding author. E-mail: singhpk67@gmail.com*

## CONTENTS

Edited version of Ashwani Kumar Madile (August 2011). Tensiometer Based Drip Irrigation Scheduling of *Capsicum annum* L. under Polyhouse. Unpublished M.Tech. (Agric. Eng.) Thesis, G. B. Pant University of Agriculture and Technology, U. S. Nagar, Pantnagar-263145, Uttarakhand, India.

## 4.1  INTRODUCTION

Efficient use of water is a key factor for precision irrigation management. Widespread efforts are being made to increase water productivity and reduce the environmental impacts of irrigation. With future water scarcity and climate change, water management will become an increasingly important issue in intensive vegetable cultivation. Vegetables play a very important role in daily diet of human beings as they are an important source of vitamins, minerals, and nutrients required for maintenance of body health. The per capita consumption of vegetables in India is very low. The available per capita vegetables in India is around 180–200 g/day against a minimum of about 300 g/day, recommended by *Indian Council of Medical Research and National Institute of Nutrition*(IICMRNIN), *Hyderabad.*

### 4.1.1  BELL PEPPER (CAPSICUM ANNUUM L.)

The capsicum or bell pepper is a vegetable crop and is mostly consumed as raw in the green mature form. The genus *Capsicum* belongs to the family Solanaceae, which is grown in several parts of the world and is believed to be native of tropical South America [61, 75]. The domesticated peppers could be broadly classified into sweet and hot types based on their level of pungency. The bell pepper (*Capsicum annuum* L.) is commonly known as sweet pepper, capsicum, or green pepper. They differ from common hot peppers in size and shape of the fruits, capsaicin content and usage. Bell pepper is one of the highly remunerative vegetables cultivated in most parts of the world especially in temperate regions of Central and South America and European countries, tropical and subtropical regions of Asian continent.

The bell pepper in India is grouped under nontraditional category of vegetables. They are mainly cultivated during *rabi* and *kharif* seasons in Karnataka, Maharashtra, Tamil Nadu, Himachal Pradesh, and Uttarakhand.

Bell pepper has attained a status of high-value crop in India in recent years and occupies a pride of place among vegetables in Indian cuisine because of its delicacy and pleasant flavor coupled with rich content of ascorbic acid and other vitamins and minerals. Nutritionally, bell peppers are rich in vitamins particularly vitamin A (180 IU) and vitamin C. Hundred grams of edible portion of capsicum provides 24 kcal of energy, 1.3 g of protein, 4.3 g of carbohydrate, and 0.3 g of fat.

A good capsicum crop can be raised on red sandy and sandy loam soils with irrigation. The crop needs more amounts of organic manure and fertilizer application. Though, crop is grown on soils with a pH range of less than 5.0 like the acid laterite soils of coastal areas and heavy rainfall *ghat* (mountain) regions to soils pH more than 8.0 like the alluvial soils of North India and black cotton soils of the Deccan Plateau. The crop performs best at a soil pH of about 6.5. Bell pepper needs good drainage and aeration, and it responds well in red loams and alluvial soils with pH of slightly acidic in nature. The capsicum crop responds well when fertilizers are given to the crop in split doses with irrigation. Under protective cover, capsicum growth is vigorous, so it is essential to manage the capsicum crop with different fertilizer and irrigation under greenhouse conditions, for quality and quantity of yield.

Basically, capsicum is a cool-season tropical crop and lacks adaptability to varied environmental conditions [74]. Capsicum crop needs ideal temperature of 26–28°C in daytime and 16–18°C in night time during the flowering stage. For vegetative growth, crop needs different temperature at different growth stages like for stem elongation the temperature ranges from 30°C in daytime to 8.5°C in night. Similarly, leaf size of this crop was optimal at 12.5°C at night. Time rate of formation of new leaves was greatest at 26°C at night. Similarly, this crop cannot withstand heavy rains during flowering or fruit set. So it is essential to provide protective cover to sweet pepper during rainy season as well as in summer season to protect the crop from excessive heat and temperature. Capsicum can be successfully grown either for fruits or for seed production using naturally ventilated (NV) poly-cum net houses for off-season cultivation in areas where temperatures do not exceed 37–38°C.

Looking into the benefits of cultivation inside greenhouse, efforts have been made to develop technology which is cost effective and economical viable for higher production. Plants grown with sufficient soil moisture throughout the crop growth stage show better growth, produce more flowers, and set more fruits. Whenever the capsicum is grown as irrigated crop, normally furrow method of irrigation is followed but the excessive irrigation may cause water logging and other fungal infection to the crop. Therefore, it is essential to manage the irrigation to the optimum level.

As the productivity of irrigated agriculture is more than twice the rainfed agriculture [57], it is very essential to bring the maximum area under irrigation. As the water is most crucial input among all inputs and required for the biological activities of the plant. Therefore, it is necessary to use this input properly and judiciously to increase the productivity of land. The cheap and easily available water resources have already been harnessed. Therefore, the scarce water is to be utilized efficiently and judiciously and for the same the solution is only to go for high-tech irrigation systems like micro irrigation system. Among sophisticated methods, drip irrigation has proved its superiority due to direct application of water and fertilizer into the vicinity of root zone. By this system, the water is withdrawn by roots with high frequency than the conventional irrigation systems. It results in better quantity and quality of the produce. Also, it saves water about 60% for different crops [23, 24]. Therefore, for management of water resources, the irrigation scheduling is essential. The irrigation scheduling depends upon the crop water requirements, which depends on the season, type of soil, and climatic conditions.

### 4.1.2  MICRO/TRICKLE/DRIP IRRIGATION

Drip irrigation was first used in mid-1960s but its wide-scale adoption commenced in 1970s when it was used on 56,000 ha. Currently, more than 6 million hectares (Mha) have been covered under micro irrigation world over. The highest coverage is in Americas (1.9 Mha) followed by Europe and Asia (1.8 Mha each), Africa (0.4 Mha), and Oceania (0.2 Mha). The top 10 countries on the basis of micro irrigated areas are: the USA, Spain, India, China, Italy, Brazil, Russia, Mexico, Saudi Arabia, and Australia. These 10 countries share about 75% of the total micro irrigated area of the world (International Commission on Irrigation and Drainage, 7th International Micro Irrigation Congress, 2006).

During the last 3 decades, micro irrigation systems have made a breakthrough in many countries around the globe owing to their capability to apply water efficiently, low labor, and energy requirements, and increase in quantity and quality of crop yield/produce. Micro irrigation encompasses drip/trickle systems, surface and subsurface drip tapes, micro-sprinklers, sprayers, micro-jets, spinners, rotors, and bubblers. Water is applied as discrete or continuous drops, or tiny streams through emitters or applicators placed along a water delivery line near the plant.

Drip or trickle irrigation refers to the frequent application of small quantities of water at low flow rates and pressures. Rather than irrigating the entire field surface, as with sprinklers, drip irrigation is capable of delivering water precisely at the plant where nearly all of the water can be used for plant growth. Because very little water spreads to the soil between the crop rows, little water is wasted in supporting surface evaporation or weed growth. The uniformity of application is not affected by wind because the water is applied at or below the ground surface. A well designed and maintained drip irrigation system can have an application efficiency of 90% [23, 24].

Indian Agriculture Today (The National Agriculture magazine, 6(11), 2006) summarized the advantage of drip irrigation such as: it has created a greater impact in terms of savings in power (358 kWH), water (53%), and labor (63%), and also in application of inputs like fertilizers (335%), and plant protection chemicals (10%). In total, there is a saving of 70% in the irrigation cost. In spite of the fact, its limitations are identified as: (1) inadequate awareness and knowledge about its application, utility, method of operation, and maintenance; (2) high initial cost, nonavailability of finance; and (3) operating problems such as nonavailability of spares/micro-tubes, high cost of components, blockage/clogging of emitters/micro-tubes, leakage due to damage of pipes/drippers, and improper after sales services.

### 4.1.3  TENSIOMETER-BASED IRRIGATION SCHEDULING

The goal of irrigation scheduling is to make the most efficient use of water and energy by applying the right amount of water to the crop at the right time, at the right place in a right manner. Proper irrigation scheduling requires a sound basis for making irrigation decisions. Methods of irrigation scheduling are based on soil water measurements, meteorological data, or monitoring plant stress. Tensiometers measure the soil water tension that can be related to the soil water content [17].

Good irrigation scheduling means making sure that water is available when the crop needs it. Scheduling maximizes irrigation efficiency by minimizing runoff and percolation losses. This often results in lower energy, water use and optimum crop yield, but can result in increased energy and water use in situations where water is not being properly managed. One of the benefits of scheduling with tensiometers is the ease of use and the immediate results. With tensiometers, users only need to look at the gauge to determine the soil moisture level with no other meters or instruments necessary. The soil water tension or suction is measured in centibars (cb) or kilopascals (kPa), which is related to the amount of water in the soil available to plants [23, 24].

Tensiometers are delicate instruments that must be handled, installed, and maintained correctly. By using the information from tensiometers, one can become more knowledgeable about the water-holding characteristics of the soil and the water needs of the plants. Tensiometers can help us to avoid plant drought stress and meet plant water needs without wasting irrigation water supplies [30].

Tensiometers continuously monitor soil water status, which is useful for practical irrigation scheduling, and are extensively used on high-value cash crops such as vegetables where low water tension is desirable. Tensiometers are ideal for sandy loam or light-textured soils. Measurement range is limited to less than one bar of tension. Tensiometers may be used in clay soils for crops that need low soil water tension for maximum yield or high crop quality.

### 4.1.4   FERTIGATION

Fertigation is defined as the application of water soluble solid or liquid (fluid) mineral fertilizers via pressurized irrigation systems, thus forming nutrient containing irrigation water. Fertigation is now the accepted method of applying most of the crops nutrition, with many growers using liquid soluble fertilizer rather than spreading granular fertilizer and waiting for the rain or sprinklers to wash the fertilizer into the root zone [23, 24].

Fertigation through drip irrigation increases fertilizer and water use efficiency (WUE) thereby minimizes the losses due to leaching and volatilization. Fertigation reduces ground water pollution and saves fertilizer and reduce application cost. Consequently, recommendations have been developed for the most suitable fertilizer formulation (including the basic nutrients NPK and microelements) according to the type of soil, physiological stage, climate, and other factors. This technology is widely used in orchard, greenhouse, and field crops. Other advantages of the fertigation are: (1) the saving of energy; (2) the flexibility time of the application (nutrients can be applied to the soil when crop or soil conditions would otherwise prohibit entry into the field with conventional equipment); (3) convenient use of compound and ready-mix nutrient solutions containing also small concentrations of micronutrients which are otherwise very difficult to apply accurately to the soil; and (4) the supply of nutrients can be more carefully regulated and monitored. When fertigation is applied through drip irrigation system, crop foliage can be kept dry thus avoiding leaf burn and delaying the development of plant pathogens. Fertigation is not optional, but is actually necessary [23, 24].

### 4.1.5   GREENHOUSE AGRICULTURAL TECHNOLOGY

A greenhouse is a framed structure made of galvanized iron pipe/MS angle/wood/bamboo and covered with transparent material or translucent material fixed to frame with grippers. Besides irrigation, it has control/ monitoring equipment, which is considered necessary for controlling environmental factors such as temperature, light, relative humidity (RH), etc., and is necessary for maximizing plant growth and productivity. Thus, the greenhouse is an enclosed area, in which crops are grown under partially or fully controlled conditions. The cladding material is of plastic (Polyethylene) film and acts like a selective radiation filter that allows solar radiation to pass through it but traps the thermal radiation emitted by the inside objects to create greenhouse effect.

Greenhouse technology protects the crop from adverse weather conditions and from attack by insects, pests, diseases; thus it helps in increasing the yield and quantity. At the same time since the inside environment remains under control, and carbon dioxide released by the plants during the night is consumed by the plants itself in the morning. Thus plants get about 8–10 times more food than the open-field conditions. The benefits, which can be derived from the green house cultivation, are:

- The yield may be 5–8 times higher than that of outdoor cultivation depending upon the type of greenhouse, type of crop, environmental control facilities.
- The crop yields are at the maximum level per unit area, per unit volume, and per unit input basis.
- High-value and high-quality crops could be grown for export markets.
- Income from the small and the marginal land holdings maintained by the farmer can be increased by producing crops meant for the export markets.
- It can be used to generate self-employment for the educated rural youth in the farm sector.

## 4.1.6   OBJECTIVE OF THIS RESEARCH STUDY

As the world population continues to increase, and more agricultural land is lost to urban development, industrialization, etc., intensive food production in greenhouses may play a more important role in food production. Furthermore, improving economic conditions in developing countries and an increasing preoccupation with health and nutrition will increase demand for high-quality food products. Through controlled climate and reduced pesticide use, greenhouses can meet this consumer demand. Foods with improved health characteristics or containing nutraceuticals (i.e., substances with pharmaceutical or health-beneficial properties that can be extracted or purified from plants) can be grown pesticide free in greenhouses.

Vegetables are fast growing, vigorous, and have most of their root system confined only on the upper layer of soil. Thus, vegetables are very sensitive to water stress. Ultimately drip irrigation systems are effective, efficient, and economically viable for irrigation of vegetables. It improves productivity and quality of produce, saves water, and economics fertilizer if used through drip (fertigation). It therefore implies that, vegetable production can become sustainable only if we use our costly and limited water and land resources with the help of modern technology to maximize crop productivity and yields and to insure higher efficiency and returns.

Keeping above points in consideration and to develop sustainable agriculture, it is important to explore the best method in a greenhouse environment by combining tensiometer-based irrigation scheduling and fertigation. Therefore, the research in this chapter was undertaken to investigate, "effects of tensiometer-based drip irrigation scheduling of capsicum (*Capsicum annum* L.) under polyhouse," with the following objectives:

1. To study the various micro environmental parameters and to estimate the crop water requirements under NVPH.
2. To study the response of different levels of irrigation and fertigation on capsicum (*Capsicum annum* L.) production.
3. To study the economic feasibility of off-season production of capsicum (*Capsicum annum* L.) under NVPH.

## 4.2 REVIEW OF LITERATURE

### 4.2.1 MICRO ENVIRONMENTAL CONDITIONS AND ESTIMATION OF WATER REQUIREMENTS

#### 4.2.1.1 MICRO ENVIRONMENTAL CONDITIONS UNDER POLYHOUSE

Graverend [26] indicated that water requirement of the crop is considerably reduced compared with conditions outside the greenhouse that is equipped with a filtering wall to allow the passage of visible light into the greenhouse and absorbs the infrared rays into a circulating liquid that stores heat for release during night.

Iqbal and Khatri [34] determined sensible convective heat transfer coefficient from two long semicircular model greenhouses. In the laboratory experiment, two 1.82 m long aluminum model of 6.03 cm and 3.17 cm diameter were used. The models were placed in a large wind tunnel of 24.4 m long and 2.4 m wide and a variable ceiling height of 1.6–2.4 m. The agriculture surface terrain (grass land and protective hedges) were simulated. The wind direction and speed were maintained normal to long axis of model. Data were presented in terms of Nusselt and Reynold Numbers.

Silveston [62] assessed the relative contribution of condensation to the total night time heat loss during cold weather operation, with a view of controlling severe condensation losses and dehumidification. Two cases were chosen for study: the first a single glazed glasshouse and the second greenhouse covered by double layer of plastic film. Results showed that significant heat loss by condensation (up to 20% of total) was expected in plastic house because of the lack of infiltration of air from outdoor.

Aviscor and Mahrer [5] designed a one-dimensional numerical model to simulate diurnal changes of the greenhouse environment. The model consisted of soil layer, vegetation layer, air layer, and a cover. The thermal radiation, sensible, latent, and conduction heat fluxes were modulated in each layer in terms of its unknown temperature and vapor pressure. An observation study was performed in order to test the ability of the model to describe the greenhouse microclimate. Good agreement was obtained between predicted and observed temperature and humidity values.

Dayan et al. [11] studied the growth conditions of tomatoes in a narrow bay, low roof Dutch greenhouse (venlo type) and in a wide bay, high roof Israeli greenhouse (sharasheret type). Various roof cover materials were also tested in Israeli greenhouse: glass polyethylene, corrugated fiber glass, and tubular polycarbonate. Differences in climatic conditions between the greenhouses were usually small. Day temperature was 1–6°C higher than the outdoor temperature. Excessively high temperature could generally be avoided by ventilation but in the Dutch greenhouse day temperature control was limited. Night temperature in all greenhouses was usually up to 1° above the outside temperature but in the polycarbonate-covered greenhouse, night temperature was higher than this. Radiation in covered greenhouses was 55–60% of the outside level and was about 10% lower than in the others. The photo synthetically active fraction of global radiation was higher inside the greenhouse than outside.

Fceilla and Cascone [16] developed a one-dimensional model to simulate the changes in thermal environment of a double-skinned greenhouse at night. The lowest night air temperature in the double-skinned structure was 1.6 times higher than that in single-skinned structure and approximately 2.5 times higher than the outside temperature.

Kittas [38] experimentally investigated the influence of an insect screen on ventilation rate in a multispan glass-covered greenhouse equipped with a continuous roof vent, located at the University of Thessaly near Volos in the coastal area of eastern Greece. Two measuring techniques were used for the determination of ventilation rate: (1) the decay rate tracer gas technique, using $N_2O$ as tracer gas, and (2) the water vapor balance technique. These measuring techniques gave similar results but the water vapor balance technique provided a better fit to the experimental data. These measuring techniques were also used to calculate the ventilation rate, and the data obtained were applied to evaluate the influence of the insect screen on ventilation rates. Results confirmed and quantified the major reduction in ventilation due to the insect screen. The wind-related coefficient was significantly decreased when an insect screen covered the vent. The results indicate that screens can reduce the discharge coefficient by 50%, and thus the ventilation rates were decreased to the same extent. For a given screen, its influence on the discharge coefficient and thus on the ventilation rate can be determined. This approach can be exploited further to determine if a given greenhouse ventilation opening design provides enough ventilation when equipped with an insect screen and to propose better vent design in order to improve greenhouse.

Kittas et al. [39] investigated greenhouse microclimate, energy savings, and rose flower transpiration during winter in a glass-covered greenhouse equipped with an aluminum thermal screen. Net radiation over and under the rose crop, heating pipe temperature, canopy temperature at 0.3 m and 0.8 m, and transpiration rate were simultaneously measured and recorded. When compared to unscreened conditions, it was found that the thermal screen provided a more homogeneous microclimate and increased the average air temperature and canopy temperature at 0.8 m by about 2.5°C and 3.0°C, respectively. The latter result was attributed to the observed increase (about 100%) in the net radiation absorbed by the canopy, along with reflection of long-wave radiation by the thermal screen. Higher canopy-to-air vapor pressure deficit were observed in the lower layer (bent shoots) under screened conditions. Canopy transpiration rate was of the same order of magnitude in both cases. Energy savings due to the screen was about 15%. These results underline that the basic effect of the studied screen on crop behavior was positive on growth, development, and sanitary conditions of rose plants.

Willits and Shuhai Li [73] compared the cooling of two NV and two fan ventilated (FV) greenhouses over two summers. The NV houses used fogging, nozzles while the FV houses had evaporative pads. In 2003, one NV house and one FV house were planted with tomato while the other house in each pair was empty. Combinations of fog, vent openings, ventilation rates, and evaporative pads were applied. No difference in fruit yield was observed. However, the fruits in the evaporative pads were slightly smaller with more defects. With evaporative pads, the temperatures (air and canopy) were lower in the FV houses than in the NV houses. Without evaporative pads, air temperatures were always higher in the FV houses. In 2004, all four houses were planted. The NV houses used all three vents plus fog and the FV houses used an airflow of 0.087 $m^3 \cdot m^{-2} \cdot s^{-1}$ plus evaporative pads. Both leaf and air temperatures in the FV houses were lower than in the NV houses, with the largest advantage occurring at the beginning of the season, declining as the season progressed. Although the temperature differences did not affect total yield over the life of the study, there was an early yield advantage to the FV houses that lasted for about a week. In addition, the NV houses produced fruits with a larger number of defects until the temperature differences between house types reached essentially zero.

Mpusia et al. [47] investigated the microclimate in a forced and a natural ventilated greenhouse; and studied quality of roses in the two greenhouses. It was concluded that the high vapor pressure deficit >3 kPa and high temperature >30°C were main causes of low quality roses (short and thin stems) produced in the forced ventilated greenhouse.

Parvej et al. [51] compared the phenological development and production potentials of two tomato varieties (Bari Tomato-3 and Ratan) under polyhouse and open-field conditions. Photo synthetically active radiation inside the polyhouse was reduced by about 40% compared to the open-field conditions while air and soil temperatures were always remained higher. From December to February, the midday air temperature under polyhouse and open-field conditions varied from 31.8°C to 39.1°C and 23.3°C to 31.1°C, respectively, indicating about 8°C higher air temperature inside polyhouse and during that time the average air temperature inside polyhouse was about 28°C, which was optimum for the growth and development of tomato crop. RH was lower inside the polyhouse compared with open-field conditions. These microclimatic variabilities inside polyhouse favored the growth and development of tomato plant through increased plant height, number of branches/plant, rate of leaf area expansion, and leaf area index (LAI) over the plants grown in open field. Flowering, fruit setting, and fruit maturity in polyhouse plants were advanced by about 3, 4, and 5 days, respectively, compared to the crop raised in open-field conditions. Polyhoused plants had higher number of flower clusters/plant, flowers/cluster, flowers/plant, fruit clusters/plant, fruits/cluster and fruits/plant, and fruit length, fruit diameter, individual fruit weight, fruit weight/plant, and fruit yield compared to open-field condition. The fruit yield obtained from the polyhouse was 81 t/ha against 57 t/ha from the open field.

Samsuri et al. [55] focused on modeling and identification of environmental climatic variables inside naturally ventilated tropical greenhouse (NVTG). The NVTG climate model is an essential tool for developing the climate control system. A real-time wireless monitoring system for measuring prototyped NVTG environmental climates for chili plantation was developed. The variables to estimate the NVTG outputs, variables namely, internal RH and internal temperature are external temperature, external humidity, irradiance, and wind speed. Modeling based on mathematical equations as well as using system identification (SI) technique is outlined. This study also focused on SI procedure namely data acquisition, identified model structure that represents NVTG using NARX and ARX model structures, parameter estimation technique, and model validation. Simulation studies revealed that linear and nonlinear polynomial models were able to model and simulate the NVTG performance.

### 4.2.1.2 ESTIMATION OF REFERENCE EVAPOTRANSPIRATION (ET₀) AND WATER REQUIREMENT

Doorenbos and Pruitt [15] utilized the same general format of American Society of Civil Engineers (ASCE) Penman with the addition of an adjustment factor and different wind coefficient.

$$ET_0 = C_p \, [WR_n + \{(1 - W) \, (0.27) \, (1 + U/100)\} \, (e_s - e_a)], \qquad (4.1)$$

$$R_n = 0.75 \, R_s - R_{nl} \qquad (4.2)$$

$$R_n = F_t \, F_{ed} \, F_{nn} \qquad (4.3)$$

$$F_{ed} = 0.34 - [0.044 \, (e_a)^{0.5}], \qquad (4.4)$$

$$F_{nn} = 0.1 + [0.9 \, (n/N)] \qquad (4.5)$$

where $ET_0$ = reference evapotranspiration (mm/day); $C_p$ = an adjustment factor dependent on maximum RH, solar radiation, daytime, wind speed, and the ratio of daytime to nighttime wind speed, $R_n$ = net radiation at the crop surface (MJ·m$^{-2}$·day$^{-1}$) defined by Equation 4.2; $R_s$ = solar radiation (mm/day); $F_t$ = function of temperature; $W$ = weighting factor dependent on temperature and altitude and equal to the slope of the SVP–temperature curve; $U$ = the 24-h wind speed at 2-m height (km/day); $e_s$ = saturation vapor pressure (kPa); $e_a$ = actual vapor pressure (kPa); $e_s$–$e_a$ = saturated vapor (kPa).

Monteith [46] developed a combination method (Penman–Monteith (PM) model) for estimating ET. The PM equation not only relates aerodynamic resistance to sensible heat and vapor transfer but also surface resistance to transfer too.

Shih [58] developed a model based on air temperature and solar radiation for estimating ET in Southern Florida for irrigation requirement prediction using an optimum ridge regression analysis. The results showed that, method gave an irrigation requirement prediction sufficiency close to that estimated by the combination method and water budget method.

Hargreaves and Samani [28] proposed a method for estimating reference ET that requires only maximum air temperatures.

Hussein and Eldraw [33] used the Food and Agricultural Organization (FAO)-Penman method with wind function for clipped grass for estimating ET of short warm-season grass when growing at its optimum temperature in Sudan Gezira. They concluded that the ET for warm-season grass and cool-season grass were equal to 1 mm/day and 0.635 mm/day, respectively.

Klocke et al. [40] compared ET rate obtained from the mini-lysimeters with prediction from combination of two energy-based models. Hargreaves [27] proposed equation for estimating ET by using climatic data. His method requires only measured values of maximum and minimum temperatures and was recommended for general use.

Manjunath et al. [45] conducted trials with okra cv. Pusa Sawani, *Capsicum annum* cv. Pusa Jwala, and cv. Pusa Rubi to study the effects of irrigation application to meet 50%, 75%, or 100% of weekly evaporative demand. Yield was affected by irrigation rate only in tomato, in which yield was highest with $(I_2)$. For okra and tomato, WUE was highest with $(I_1)$; WUE of *Capsicum annum* L. was not affected by irrigation rate.

Chiew et al. [10] compared the reference evapotranspiration $(ET_0)$ estimated using the PM and FAO-24 methods and class A pan (CAP) data of 16 Australian locations with a wide range of climatic conditions. The analysis indicated that the FAO-24-PM $ET_0$ estimates were 20–40% higher than the PM estimates; however, the FAO-24 radiation and PM methods gave similar daily $ET_0$ estimates. FAO-24-Blaney–Criddle method, which uses only the temperature data, gives similar monthly $ET_0$ estimates as PM and is, therefore, adequate for application in long-term $ET_0$ estimates. The comparison also showed that there was a satisfactory correlation between CAP data and PM $ET_0$ for total evaporations over three or more days. However, the pan coefficient was dependent on local climate and physical conditions and it was suggested that it should be determined by comparing the Penman data with either the PM or FAO-24-radiation $ET_0$ estimates.

Pereira et al. [3, 52, 64] estimated standard concepts of potential evaporation $(E_p)$ and equilibrium evaporation by introduction of climatic resistance provided with aerodynamic and surface resistance. The results showed that approaches for computing $ET_c$ using $K_c \times ET_0$ ($K_c$ is crop coefficient) are compatible with direct estimation of $ET_c$ using PM formulation or other resistance-based approaches.

Samani [54] introduced a procedure to estimate solar radiation and subsequently reference crop ET using minimum climatological data. A modification to an original equation that used maximum and minimum temperature to estimate solar radiation and reference crop ET was described. The proposed modification allowed for the correction of errors associated with indirect climatological parameters affecting the local temperature range and it also improved the accuracy of estimates of solar radiation from temperature.

Watanbe et al. [72] estimated the ET values for irrigation planning system in Djibouti based on the monthly average values of air temperature and RH in the area. The relationship between the daily average air temperature and altitude showed that the higher the altitude, the lower the daily air temperature. From this principle, it was concluded that the amount of daily water consumption is also based on the altitude.

Klosowski and Lunardi [41, 42] conducted a greenhouse experiment to determine the ET/crop water use and crop coefficient of red pepper at several growth stages. The crop ET was 239.5 mm in a 198-day vegetable cycle, with a daily average of 1.5 mm/day. The greater water consumption was 138.5 mm at the intermediate stage. The crop coefficient values were increased until the intermediate growth stage with an average of 0.8.

Prenger et al. [53] compared four models that represented the progression from outdoor, "big leaf" estimates to indoor, greenhouse-specific formulations. Each of the models was algebraically reduced to common and consistent terms. The data from a lysimeter study were used to compare ET combination models using ET rates of red maple tree (*Acer rubrum* var. Red Sunset) grown in a controlled-environment greenhouse. The measured ET was compared with two empirical climatic factors (solar irradiance and vapor pressure deficit) and with calculated ET based on four ET models: (1) Penman, (2) PM, (3) Stanghellini, and (4) Fynn. A measure of the model performance was evaluated by the Nash–Sutcliffe $R^2$, or model efficiency. The relationship between measured and calculated ET for the Stanghellini model had a model efficiency of $R^2 = 0.872$, while the other models yielded correlation coefficients of $R^2 = 0.214$, $R^2 = 0.481$, $R^2 = -0.848$ for Penman, PM, and Fynn, respectively. The differences between the models revealed the importance of the LAI factor and a sub-model for irradiance in the plant canopy. Based on the coefficient of

determination, vapor pressure deficit alone yielded a good linear correlation to estimate ET ($R^2$ = 0.884), while irradiance alone resulted in
$R^2$ = 0.652.

Blanco and Folegatti [6] determined the evapotranspiration ($ET_c$) and
crop coefficient ($K_c$) of cucumber in a greenhouse during the winter–spring
season in Piracicaba-SP-Brazil. Crop was irrigated with water of three
different levels of salinity: S1 = 1.5, S2 = 3.1, and S3 = 5.2 dS/m. Irrigation was performed when soil matric potential reached −30 kPa, which
was determined by the mean matric potential at 0.15-m and 0.30-m soil
depths, and the depth of irrigation was calculated from a reduced-evap-
oration pan. The $ET_c$ and $K_c$ were reduced linearly by the salinity of the
irrigation water with reduction in $ET_c$ of 4.6% per unit increase of salinity.
Measured $K_c$ values for S1 were very close to the estimated values; thus
the combined use of tensiometers and evaporation pan was found to be
adequate for irrigation management in greenhouse.

Mpusia et al. [47] investigated the crop water requirement for roses
grown outdoor and in greenhouse (commercial multi-span) in Naivasha,
Kenya. The actual ET in the greenhouse was estimated by water balance
(WB) method in hydroponics. The difference between the water application depth and drainage gave the actual ET (mm/day). For outdoor conditions, the actual ET with PM equation in the greenhouse was 65% of actual
ET outdoor.

Gomez et al. [20] evaluated the Priestley–Taylor (PT) model to estimate the real evapotranspiration ($ET_{real}$) of a drip-irrigated greenhouse
tomato (*Lycopersicon esculentum* Mill.) crop. The net radiation incorpo-
rated in the PT model was estimated using meteorological variables. For
this experiment, an automatic weather station (AWS) was installed inside
the greenhouse to measure solar radiation ($R_{gi}$), net radiation ($R_n$), air
temperature ($T_a$), and RH. Another AWS was installed for a grass cover to
measure atmospheric conditions outside the greenhouse. The PT model
was evaluated using the $ET_{real}$ obtained from the WB method. In this case,
values of $ET_{real}$ by PT model were calculated using: (a) $R_{gi}$ and soil heat
flux ($G$) = 0; (b) Rgi and $G \neq 0$; (c) solar radiation measured outside the
greenhouse ($R_{ge}$) and $G$ = 0; and (d) $R_{ge}$ and $G \neq 0$. For these four cases,
results indicated that PT model was able to compute $ET_{real}$ with errors
less than 5%. Also, Rn was calculated with a relative absolute error and
a mean deviation lower than 6% and 0.07 mm·$d^{-1}$, respectively, using

$R_{gi}$ or $R_{ge}$. Daily soil heat flux values equal to zero did not affect the calculation of $ET_{real}$ values. Thus, the PT model evaluated in this study can be used for scheduling irrigation for a greenhouse tomato crop, using internal measurements of air temperature and RH, and external measurements of solar radiation. In this case, PT model predicted the $ET_{real}$ with an error of 6.1%.

M. Casanova et al. [43] evaluated five methods to estimate crop ET in greenhouse conditions and the performance was compared; the ET was directly determined from WB measurements ($ET_{lys}$), in an irrigated lettuce (*Lactuca sativa* L.) crop during 9 weeks. Daily values of the reference evapotranspiration ($ET_o$) were compared from CAP, Piche atmometers (ATM), Andersson evaporimeters (ANE), FAO-Radiation (FRE) and FAO-PM (PME) equations. The methods showed similar temporal variations in the order ANE < CAP < FRE < PME < ATM. Furthemore, $ET_o$ had a clear correlation with solar radiation. Crop coefficients ($K_c = ET_{lys}/ET_o$) varied somewhat amongst the methods, but trends were identified for two periods: in the first week, the overall mean $K_c$ was 0.3 ($\pm 0.1$) and in 2–9 weeks, the average was 0.6 ($\pm 0.3$). The greenhouse values of $K_c$ were lower than those generally adopted for lettuce in field conditions. In terms of irrigation design, crop ET can be estimated by the methods in this study, on the condition that the appropriate crop coefficients are applied. The fact that ANE showed values closest to those of $ET_{lys}$, along with cost and management convenience, makes it an advantageous alternative compared to the other methods.

Takakura et al. [68] estimated ET in a greenhouse with the energy balance equation, and an instrument was developed to collect data for this purpose. The values estimated by this method were in good agreement with the measured data. It was shown that the net solar radiation term was the largest and cannot be neglected, and that long-wave radiation exchange had a relatively small effect. As usual, soil heat flux can be neglected but the sensible heat transfer term cannot be neglected since the maximum of the possible range of values is large and significant. It was concluded that the method was simple and suitable for irrigation control in greenhouses. It was also concluded that normal radiation sensor measurements on a horizontal surface are not adequate for measuring radiation received by a plant canopy in a single-span greenhouse.

## 4.2.2   IRRIGATION AND FERTIGATION SCHEDULING

### 4.2.2.1   PERFORMANCE EVALUATION OF TENSIOMETER-BASED IRRIGATION

Abbott et al. [1] investigated the effects on yield and radial fruit cracking of two media (soil in beds and soilless medium in bags) and two drip irrigation frequencies (once and four times daily) on four greenhouse tomato (*Lycopersicum esculentum* L. Mill.) cultivars. For plants grown in soilless medium, two tensiometer-controlled, drip irrigation scheduling methods were compared. 'Michigan' and 'Ohio' hybrid fruits cracked significantly more than the three remaining cultivars, but did not differ in production of total and number one fruit. The amount and severity of fruit cracking was least from the soilless, bag-cultured plants. Total mean fruit weight was greatest from soil-grown plants. Although no differences in cracking occurred in the fruit from soilless, bag-cultured plants, those whose irrigation was based on soilless medium tensiometer readings produced lower total mean fruit weight than those whose irrigation was based on soil tensiometer readings. Number and weight of defective fruit was lowest from plants grown in soilless medium and whose irrigation was based on soil tensiometer readings, and greatest from soil-grown plants. Fruit cracking was reduced by increasing the irrigation frequency from 1 to 4 times daily.

Smajstrla (1990) studied effects of tensiometer-based irrigation scheduling with pan evaporation-based irrigation on tomatoes (*Lycopersicon esculentum* Mill.), that were grown on Arredondo fine sand soil using drip irrigation, polyethylene mulch. Irrigations were scheduled with tensiometer and pan evaporation to determine the effects of these irrigation scheduling practices on irrigation water requirements and fruit yield. Irrigation applications were greater with the 10 cb than the 15 cb treatment. Yields of extra-large and total fruit were significantly greater with all treatments than the control. Marketable tomato yield with the tensiometer-based treatments was not significantly different than with the pan evaporation treatment.

Chartzoulakis et al. [9] determined the water consumptive use of sweet peppers grown in an unheated greenhouse by using tensiometers. Water supplied to restore 85% of maximum evapotranspiration ($ET_m$) had no effect on plant growth and fruit yield; with irrigation to restore 65% or 40% of ($ET_m$) plant growth was declined and the total yield was reduced by 26% and 47%, respectively. Both fruit numbers per plant and fruit size were affected by the amount of water applied.

Kruger et al. (1999) compared irrigation scheduling based upon tensiometer measurement and climatic WB model to the nonirrigated one. The influence on fruiting response, the impact on soil moisture tension, mineral nitrogen content in the soil, and nitrogen content in the leaves were investigated. During all the years, irrigated plants had significantly higher yields than the nonirrigated ones. The mean fruit weight was also increased by irrigation. Optimization of irrigation was best achieved in both varieties when the climatic WB model was used to schedule irrigation. During dry periods, soil moisture tension under nonirrigated strawberries was increased to values above 300 kPa at 20-cm soil depth. In a year, with dry conditions and high ET, maximum values above 700 kPa were reached. Differences in soil moisture were observed between the two irrigation schedulings within and over the years of the experiment. The hydraulic gradients calculated from tensiometer measurements showed that in the periods with irrigation, percolation did not occur below 40-cm soil depth. Therefore, leaching of mineral nitrogen out of the rooted zone of strawberries could be excluded for both irrigation schedules.

Olczyk et al. [50] investigated that irrigation is a critical factor for winter vegetables grown on calcareous soil in Southern Florida, where the soil is characterized by low nutrient and low water-holding capacity. Traditional approaches to irrigation are based on stage of the growth or visual estimation of soil moisture. This often leads to over irrigation or under irrigation of the crops. Under irrigation may reduce the yield and quality, while over irrigation may lead to leaching of nutrients from root zone, contributing to pollution. A demonstration with using tensiometers to schedule irrigation was conducted in a commercial field in Homestead area to demonstrate the usefulness of tensiometers in such applications. The frequency of irrigation events and amount of water depth were based on tensiometer readings compared to the grower's irrigation scheduling. Data included total number, weight, and quality of fruits. Results showed that reduction of irrigation water did not decrease total marketable yield. Tensiometer with proper calibration, installation, and maintenance can be successfully used for scheduling irrigation.

Yuvan et al. (2000) studied the water consumption of tomato and the relationship between ET and water surface evaporation measured with

CAP, where the soil water potential in root zone is controlled by tensiometer. It was found that there is not significant water flux at 0.75-m depth when soil water potential is kept higher than −20 kPa at 15 cm, and around −20 kPa at 60 cm. The accumulative value of 517 mm ET at any time approximates accumulative value of water surface evaporation measured using pan.

Hoppula et al. [32] studied three different drip irrigation thresholds at −150 hPa, −300 hPa and −600 hPa maintained by four tensiometers in each treatment, in combination with either fertilizer in fertigation or broadcast in Southwest Finland. Only a few significant differences were observed in the effects of various treatments on flowering, yield, size of berries, or vegetative growth of blackcurrant cv. Mortti. The soluble solids content (Brix) was highest when using fertigation and irrigation threshold of −150 hPa. Based on this study and earlier studies, the optimal soil moisture measured with tensiometer in sandy soils is at least −270 hPa or even more moist. The soil moisture effect depends on soil nutrient status.

Wang and Zhang [71] conducted experiment in greenhouse tomato plants which was subjected to five subsurface irrigation treatments with maximum allowable depletion (MAD) of 210, 216, 225, 240, and 263 kPa of soil water potential, respectively. The long-term effect of subsurface irrigation scheduling on soil neutral phosphatase activities at five soil depths (0–10, 10–20, 20–30, 30–40, and 40–60 cm) were investigated. Results showed that subsurface irrigation could enhance soil neutral phosphatase activity in the treatment with higher irrigation MAD. Neutral phosphatase activities were higher in the topsoil than subsoil, with heightened phosphatase activities at a depth of 10–20 cm in the soil profile in subsurface irrigation. Neutral phosphatase activity presented significantly positive linear relationships with available phosphorus (P) and contributed to the increase of available P in soil. Irrigation management could be applied to adjust phosphatase activity and MAD of 210–216 kPa is an advisable subsurface irrigation scheduling to heighten phosphatase activity, thereby contributing to higher P availability in soil.

Gonzalez et al. [21] studied three well-watered vegetable crops with soil–water matric potential (SMP) values between −20 and −30 kPa throughout most of the respective growth cycles. These values avoid water deficits in Mediterranean greenhouse vegetable crops. The watermelon under (deficit irrigation strategy) regulated deficit irrigation

(RDI) presented similar SMP to the well-watered crop, except during the flowering period when it reached values of −50 to −60 kPa, which are similar to, or slightly lower than those recommended to prevent water deficits for cucurbitaceae crops. The autumn–winter and spring cycles of green bean under RDI presented progressively lower SMP values from the vegetative phase to the first fruit setting than the well-watered crops, reaching minimum SMP values of around −55 kPa for the autumn–winter cycle and of −75 kPa for the spring one. These minimum SMP values are similar for the autumn–winter cycle and lower for the spring cycle than those recommended to avoid water deficits in green bean crops grown in medium-fine-textured soils. Overall, mild water deficits during flowering stage of watermelon and green bean crops grown did not improve the final fruit number or yield. In the two spring cycles (watermelon and green bean), the RDI strategy reduced the aboveground biomass and yield, whereas in the autumn–winter green bean cycle the RDI strategy reduced the vegetative biomass but did not affect yield. SMP threshold values can, however, be used by growers as a tool for controlling the equilibrium between the vegetative and reproductive growth of greenhouse soil-grown crops

Zhang et al. [77] studied to select suitable indicator for scheduling the irrigation of jujube (*Ziziphus jujuba* Mill.) grown in Loess Plateau. The relationships between plant-based indicators and soil matrix potential as well as meteorological factors of jujube under deficit irrigation compared with well-irrigation were determined. The results showed that maximum daily trunk shrinkage was increased and maximum daily trunk diameter, gas conductance, and midday leaf water potential were decreased in response to higher and lower soil matrix potential, respectively. However, the maximum daily trunk shrinkage signal intensity to noise ratio was highest in response to higher and lower soil matrix potential. Besides, the maximum daily trunk shrinkage was correlated well with reference ET and vapor pressure deficit ($R^2 = 0.702$ and $0.605$, respectively). When the soil water potential was greater than −25 kPa or less than −40 kPa, maximum daily trunk shrinkage values showed an increasing trend, suggesting that Jujube might be subjected to water stress. Based on this, the suitable soil water potential values of pear-jujube during anthesis and setting periods were between −40 and −25 kPa and the values can help to conduct precision irrigation of jujube in the Loess Plateau.

## 4.2.2.2   RESPONSE OF DIFFERENT IRRIGATION METHODS AND FERTIGATION ON HORTICULTURAL CROPS

Sumarna and Kusandriani [67] conducted experiments in screen house to study the response of plant to the amount of water supply that gave better influence on growth and yield. The results revealed that there was significant interaction between the amount of water supply and cultivar on the plant height of sweet pepper at 45 days after transplanting (DAT) and 60 DAT. An appropriate crop water requirement for growth and development of sweet pepper at vegetative stage was 200 ml for flowering and 400 ml for fruit setting of sweet pepper during generating stage.

Nicole and George [49] studied N-fertilization management for drip-irrigated bell pepper (cv. Camelot bell pepper) that was fertigated and grown on an Arrendondo fine sandy soil using polyethylene and fumigated beds. Portions of total season N (0%, 30%, 70%, and 100%) were applied to the soil at bed formation. The remaining N was injected weekly into the drip irrigation system. Total N application treatments were 85, 170, 255, or 340 kg/ha. Early and total-season marketable fruit yields were increased quadratically with N rate. Preplant N fertilizer proportion did not influence the early yield, but second and total-season marketable fruit yields were decreased linearly as preplant fertilizer proportion increased. Whole-leaf N concentrations with all treatments were higher than critical values (>40 g/kg) throughout the season.

Deolankar and Firake [13] studied the effects of solid soluble fertilizers (SSF) on growth and yield of chili in a clay loam soil. It was found that the water requirement of chili under drip irrigation was 476.55 mm. Drip irrigation saved 58.2% water compared with the control. The WUE was also higher (20.14 kg/ha-mm) in the 125% recommended dose of SSF as against 5.82 kg/ha-mm in control. The application of recommended dose of P and K to soil at transplanting and N as urea through drip proved better than control in all parameters.

Goswami et al. [22] conducted two field experiments inside the plastic houses to evaluate the bell pepper response to nitrogen fertigation. The results indicated that the yield and marketable number of fruits in both seasons ware increased with the addition of nitrogen and also the peak nitrogen requirement and utilization was during the period of the maximum growth rate (90–159 DAP).

Hills and Brenes [31] evaluated effects of four different types of drip tapes on secondary effluent from an activated sludge wastewater treatment plant. Additional treatment within the irrigation system included sand media filtration, continuous chlorination with a free chlorine residual concentration of 0.4 mg/L through the drip tapes. During the first 2-month phase of continuous operation, none of the drip tape suffered flow reductions of more than 5%. During the 6-month second phase, better performance drip tapes was used for additional assessment. The statistical uniformity of emitter flow in four drip tapes ranged between 92.7% and 98% (mean value of 94.8%) during the second phase. This study indicated that drip tape technology has been significantly improved in recent years. Despite its low operating pressure and emission rate, relative to other micro-emitters, drip tape appears to be a suitable product for use with activated sludge secondary effluent.

Shinde et al. [59] evaluated micro irrigation systems and mulching for summer chili production at Agronomy Farm, Dr. Balasaheb Sawant Konkan Krishi Vidyapeeth, Dapoli. The study included different irrigation methods (check basin, micro-jets, and drip irrigation systems) with varying levels of water depth. The study revealed that micro-jet with 50% irrigation water with mulch recorded superiority regarding average number of fruits per plant. The drip with 25% irrigation with mulch recorded the highest irrigation water use efficiency (IWUE), compared to the lowest with the conventional check basin method.

Sarker et al. [56] studied response of chili to integrated fertilizer management in Bangladesh. Field experiments were conducted to study response of chili cv. Balijhuri to different doses of N, P, K, S, and Zn under rain-fed conditions. The study concluded that the effects of nitrogen were highly pronounced. Similarly, P, K, S, and Zn benefitted the number of fruits per plant and fruit yield. The economic analysis recommended that the treatment combination 100:90:90:20:2 kg/ha of $N:P_2O_5:K_2O:S:Zn$ is more suitable option for fresh yield of chili and getting the highest marginal rate of return.

Harmanto et al. [29] experimented on four different levels of drip fertigated irrigation equivalent to 100%, 75%, 50%, and 25% of crop evapotranspiration ($ET_c$), based on PM method, to study the effects on crop growth, crop yield, and water productivity. Tomato (*Lycopersicon esculentum*, var. Troy 489) plants were grown in a poly-net greenhouse. Results were compared with the open cultivation system. Two modes of irrigation application namely continuous and intermittent were used. The distribution uniformity, emitter flow rate, and pressure head were used to evaluate the performance of drip irrigation system with emitters of 2, 4, 6, and 8 Lph discharge. The results revealed that the optimum water requirement of tomato was around 75% of the $ET_c$. Based on this, the actual irrigation water for tomato in tropical greenhouse could be recommended between 4.1 and 5.6 mm/day or equivalent to 0.3–0.4 l/plant/day. Statistically, the effect of depth of water application on the crop growth, yield, and irrigation water productivity was significant, while the irrigation mode did not show any effect on the crop performance. Drip irrigation at 75% of $ET_c$ provided the maximum crop yield and irrigation water productivity. Based on the observed climatic data inside the greenhouse, the calculated $ET_c$ matched the 75–80% of the $ET_c$ computed with the climatic parameters observed in the open environment. The distribution uniformity dropped from 93.4 to 90.6%. The emitter flow rate was also dropped by about 5–10% over the experimental period. This is due to clogging caused by minerals of fertilizer and algae in the emitters. It was recommended that the cleaning of irrigation equipments (pipe and emitter) should be done at least once during the entire cultivation period.

Khan et al. [36] reported the effects of different irrigation intervals on bell pepper. The study consisted of irrigation after 3-day interval, 6-day interval, and 9-day interval and control. The study revealed that maximum seedling survival of 93% was observed in plots with 3 days of irrigation interval followed by 85% in treatments with 6-days interval. It was observed that the maximum plant height, number of leaves/plant, and leaf area were significantly higher in treatment having irrigation after 3 days of intervals compared to other treatments. The reproductive parameters of bell pepper (number of flowers per plant, number of fruits/plant and fruit weight/plant) was maximum in irrigation with 3 days of interval. The crop growth and production of fruit also gave highest yield with irrigation with 3 days of intervals. It was concluded that 3-day irrigation interval was better compared to other treatments.

Bonachela et al. [7] elucidated irrigation scheduling based on the daily historical crop evapotranspiration $(ET_h)$ data and they experimentally assessed the major soil-grown greenhouse horticultural crops in Almeria coast in order to improve irrigation efficiency. Overall, the simulated seasonal $ET_h$ values for different crop cycles from 41 greenhouses were not significantly different from the corresponding values of real-time crop evapotranspiration $(ET_c)$. Additionally, for the main greenhouse crops on the Almeria coast, the simulated values of the maximum cumulative soil water deficit in each of the 15 consecutive growth cycles (1988–2002) were determined using simple soil WB comparing daily $ET_h$ and $ET_c$ values to schedule irrigation. In most cases, no soil water deficits affecting greenhouse crop productivity were detected. The response of five greenhouse crops to water applications scheduled with daily estimates of $ET_h$ and $ET_c$ were evaluated. In tomato, fruit yield did not differ statistically between irrigation treatments, but the spring green bean irrigated using the $ET_h$ data presented lower yield than that irrigated using the $ET_c$ data. In the remaining experiments, the irrigation management method based on $ET_h$ data was modified to consider the standard deviation of the inter-annual greenhouse reference ET. No differences between irrigation treatments were found for productivity of pepper, zucchini, and melon crops.

Brar et al. [8] determined the effects of three levels of irrigation and two levels of light in open and poly house conditions on transpiration in capsicum under polyhouse. The light intensity was increased with the advancement of day, attaining a peak at noon, and there was heating of air as well as the crop canopy, which resulted in higher leaf temperatures causing transpiration to increase. Hence, the leaf temperature was positively correlated with the light intensity. There was not much difference in leaf temperature among the main irrigation treatments and sub-treatments. Highest leaf temperature was observed at 2 DAT in all treatments. It was observed that outside the polyhouse, leaf transpiration was less than that inside the polyhouse. Under different irrigation treatments with increase in irrigation levels, leaf transpiration also increased. The transpiration under irrigation remained high and kept canopy cool.

Khurana et al. [37] evaluated the effects of nitrogen in split doses on growth and yield of chili. Nitrogen was applied in form of urea in split dose at an interval of 30 days (60 kg/ha in two split doses in March and April, 90 kg/ha in three split doses in March–May, 120 kg/ha in four split doses from March to June, 150 kg/ha in five split doses in March–July, and 180 kg/ha in six split doses from March–August). Phosphorus and potassium were applied as per recommendations. The study revealed that application of nitrogen had significant effect on crop. A minimum yield of 9480 kg/ha was obtained when plants were not applied with nitrogen. It recorded a linear increase in yield of red ripe fruits with every additional dose of nitrogen and the highest yield of 1951 kg/ha was recorded with nitrogen per hectare being statistically at par with 180 kg (1883 kg/ha) and 120 kg (1858 kg/ha) nitrogen at 150 kg/ha. The increasing yield at the higher nitrogen level was possibly due to increase in synthesis of photosynthetic because of increased vegetative growth reflected in terms of plant height and number of branches. It was concluded that application of nitrogen at 120 kg/ha was beneficial if applied in four equal split doses for improving growth and yield of chili.

Mahajan and Singh [44] conducted a 2-year study to investigate the effects of irrigation and fertigation on greenhouse tomato. Drip irrigation at $0.5 \times E_{pan}$ along with fertigation of 100% recommended nitrogen resulted in an increase in fruit yield by 59.5% over control (recommended practices) inside the greenhouse and by 116.2% over control (recommended practices) outside the greenhouse, respectively. The drip irrigation at $0.5 \times E_{pan}$ irrespective of fertigation treatments gave a saving of 48.1% of irrigation water and resulted in 51.7% higher fruit yield as compared to recommended practices inside the greenhouse. Total root length was higher in drip-irrigated crop as compared with surface irrigated crop. Greenhouse tomato fruits were superior to fruits from open-field conditions, total soluble solids (TSS) content, ascorbic acid content, and pH. Further, drip irrigation in greenhouse crop caused significant improvement in all the quality characteristics.

Tumbare et al. [70] studied the effects of planting techniques and fertigation on the productivity and nutrient uptake of summer chili. The study concluded that the uptake of nitrogen and phosphorus was significantly higher in triangular planting techniques. Higher uptake of nitrogen, phosphorus and potassium was observed with a recommended dose of fertilizer at 30%, 40%, 10%, 20% at 1–2, 3–4, 6–9, and 10–15 weeks after transplanting.

Thorat et al. [69] studied the effects of protective cover (shed-net) with different levels of fertilizer and irrigation on growth and yield of capsicum. It was found that statistical superior values for number of flowers, number of fruits, average yield were recorded by treatment combination with 100%-pan evaporation irrigation level and 120% recommended dose of fertilizer. The treatment combination gave maximum net income of 767,000 Rs./ha. The treatment showed superiority over other treatments in terms B:C ratio The study concluded that the 100%-pan evaporation and 120% of recommended dose of fertilizer was superior over other treatments.

Demirtas and Ayas [12] determined the effects of deficit irrigation on yield for pepper under unheated greenhouse condition in Bursa, Turkey. In the study, irrigation depth was based on 100%, 75%, 50%, 25%, and 0% (as control) of CAP evaporation ($K1_{cp}$ 1.00, $K2_{cp}$ 0.75, $K3_{cp}$ 0.50, $K4_{cp}$ 0.25, $K5_{cp}$ 0.00 (control)) corresponding to 2-day irrigation frequency. Irrigation water depth to crops ranged from 65 to 724 mm, and water consumption ranged from 115 to 740 mm. The effects of irrigation water level on the yield, fruit height, diameter and weight, and dry matter ratio were significant. The highest yields were 24 and 19 t/ha for the $K1_{cp}$ and $K2_{cp}$ treatments, respectively. Crop yield response factor ($k_y$) was 1.07. The highest values for WUE and IWUE were 3.13 and 3.39 kg/mm for the $K2_{cp}$ treatment. Under the conditions of scarcity of water resources, it can be recommended that $K2_{cp}$ treatment is most suitable as a water application level in pepper under drip irrigation and unheated greenhouse condition [76].

Gercek et al. [19] studied the effects of water pillow irrigation method (a new alternative method to furrow irrigation) on the yield and WUE of hot pepper in a semi-arid region. Although the plants were grown under different irrigation methods and interval conditions, there were no statistical differences in yield and biomass of hot pepper plants between FI and water pillow (WP) treatments. WUE and IWUE values were significantly increased with the application of WP methods. The highest WUE and IWUE values were obtained from WP irrigation with 11-days interval treatment in both years. The study concluded that WP method can save water and increase the yield in semi-arid regions where climatic conditions require repeated irrigation.

Graber et al. [25] investigated the impact of additions (1–5% by weight) of a nutrient-poor, wood-derived biochar on *Capsicum annuum* L. and *Lycopersicum esculentum* Mill. plant development and productivity in a coconut fiber-tuff growing mix under optimal fertigation conditions. Pepper plant development in the biochar-treated pots was significantly enhanced as compared with the unamended control plots. This was reflected by a system-wide increase in most measured plant parameters (leaf area, canopy dry weight, number of nodes, and yields of buds, flowers, and fruit).

Samsuri et al. [55] found that the Malaysian production of vegetables, flowers, and spices in greenhouses has been experiencing accelerated growth. Most of the greenhouses were allowing automatic control of water, fertilizers, and climate systems. The fertigation demands the use of soluble fertilizers and pumping and injection systems to introduce the fertilizers directly into the irrigation system. Fertigation provides adequate nutrient quantity and concentration to the need through the growing season of the crop. Consequently, the design provides control of fertilizer mixing process using precise proportional pump injector flow rate with control time-based injection at predecided electrical conductivity value followed by plant nutrient uptake rate on time-based irrigation system. Planning the irrigation system and nutrient supply to the crops according to their physiological stage of development, and consideration of the soil and climate characteristics, result in high yields and high-quality crops.

## 4.2.3  ECONOMIC FEASIBILITY OF CAPSICUM UNDER POLYHOUSE

Sondge et al. [65] determined the response and economic analysis of water and nitrogen to chilli. The results revealed the irrigation scheduling at 0.75 IW/CPE with 60-mm depth coupled with 112.5 kg of N/ha recorded significantly highest yield of green fruits. The relationship was quadratics between yield versus irrigation and nitrogen fertigation. Authors recorded highest yield of 8808 kg/ha with the irrigation depth of 597.7 mm and fertigation of 129.84 N-kg/ha. The economically optimum levels of water and nitrogen under upraised price spectrum ranged from 587.5 to 593.60 mm and 126.87 to 129.8 kg/ha, respectively. The gross returns ranged over 17,460–20,198 Rs./ha.

Desai [14] reported that the application of 1 t/ha of vermicompost in capsicum resulted in lower yield as compared to chemical treated plot. However, net profit was higher in vermicompost treated plots due to less total input cost.

Shinde and Firake [60] studied the economics of summer chilli production with mulching and micro irrigation in clay soil. Among all treatments of micro irrigation systems, the maximum gross monetary returns were observed in rotary micro-sprinkler (142,324 Rs./ha) and that of minimum in strip tape (127,512 Rs./ha). Amongst micro irrigation treatments, the maximum seasonal cost of production was in rotary micro-sprinkler (62,530 Rs./ha), which was 165% more over control and was minimum in cane wall (45,733 Rs./ha), which was 93.89% over control due to its initial investment. It was found that amongst micro irrigation systems, the cane wall drip tape was most economical with benefit-cost ratio (BCR) of 2.85:1 and extra income of 42,164 Rs./ha, over bordered layout (control). However, the maximum net extra income was observed in stationary micro-sprinkler with BCR of 2.74:1. Amongst different mulches, the sugarcane trash mulch was effective due to highest BCR of 3.80:1 with the net extra income of 30,114 Rs./ha over control. The highest seasonal cost of production (62,530 Rs./ha) was rotary micro-sprinkler due to high cost of system.

Jeevansab [35] obtained the highest net returns (273,038 Rs.) and gross returns (429,600 Rs.) in capsicum crop with the treatment combination of VC + RDF under polyhouse condition.

Sriharsha [66] recorded highest net returns of 255,267 Rs. and gross returns of 406,738 Rs. in tomato production with the treatment combination of vermicompost + RDF in low cost polyhouse condition.

Shinde et al. [59] studied the effects of micro irrigation systems and nitrogen levels on growth and yield of chilli under clay loam soil. It was observed that micro-jet with 50% irrigation water with mulch recorded the highest gross monetary returns (244,080 Rs./ha) followed by recommended check basin irrigation with mulch (206,880 Rs./ha) and micro-jet with 75% of irrigation water with mulch (201,000 Rs./ha), respectively. The net profit was increased by 61,512.70 Rs./ha in micro-jet with 50% irrigation due to application of mulch at 7 t/ha, as compared with micro-jet without mulch. In case of micro irrigation system, the highest BCR (1.7:1) was noticed due to micro-jet with 50% irrigation water with mulch followed by check basin with mulch (1.58:1). The lowest BCR of 0.80:1 was observed in drip irrigation treatments with 40% irrigation water.

Singh et al. [63] estimated that the cost of construction of greenhouse is around 500 Rs./m$^2$. The life of structure was considered as 20 years whereas life of glazing material is about 4 years. The net income was 7 and 8 Rs./m$^2$ per season for production of tomato and capsicum, respectively, inside greenhouse. Cultivation of tomato and capsicum under greenhouse will not only help in getting higher productively but also fetch better returns because of the premium price for the excellent quality.

Gajanana et al. [18] reported that the cost of establishing Fan and Pad Polyhouse (FPPH) was almost double the amount required for NVPH. The cost of establishment of FPPH was 437,000 as against 220,500 Rs. in case of NVPH. The cost/m$^2$ area was 613 for NVPH and 1215 Rs. for FPPH. The major additional costs contributing to higher establishment cost of FPPH were due to polyhouse structure, fan, pad, pad motor, and electrification. NVPH gave net returns of 136.5 Rs./m$^2$, whereas FPPH registered losses were 160.02 Rs./m$^2$, due to the heavy investment made on structures, fan, pad, and electricity in case of FPPH resulting in higher cost of cultivation without any additional benefit in terms of yield or quality of flowers. Hence, it may be desirable to cultivate gerbera under NV low cost polyhouses under Bangalore (India) conditions rather than under costly structures with fan and pad cooling system.

Anitha et al. [4] reported that the highest net return (26,050 Rs./ha) was obtained with vermicompost at 10 t/ha along with full recommended dose of fertilizer. Cost of cultivation was higher in all the organic manure applied plots compared with control. Among the integrated nutrient management treatments, application of vermicompost at 10 t/ha in combination with NPK at 70:25:25 kg/ha, which recorded the highest BCR (1.89) and was most economical integrated nutrient management practice for oriental pickling melon.

Agarwal and Satapathy [2] investigated that the low cost polyhouses may serve as profitable enterprise for farmers of hilly region to grow off-season vegetables. Since such greenhouse uses locally available bamboo and other construction material, the cost of frames is lower as compared with steel framed structure. Although the life expectancy of the structure is lower compared with steel structure, bamboo is easy to replace and expertise is locally available. Such low cost technology needs popularization and demonstration in farmer's field for higher adaptability. However, cropping sequence needs to be established for such polyhouses to maximize the profit.

Murthy et al. [48] studied the economic viability of production of capsicum and tomato in a NVPH of medium cost category with drip irrigation system. Data were generated by cost accounting method for estimating the feasibility of production and was analyzed by using project evaluation methods, like Pay Back Period (PBP), BCR, Net Present Value (NPV) and Internal Rate of Return (IRR). Cultivation of capsicum in a polyhouse was highly feasible with higher values of NPV (323,145 Rs./500 m$^2$), BCR (1.80) and IRR (53.7%) with PBP of less than 2 years. Breakeven price for capsicum production in a polyhouse (11.80 Rs./kg) was lesser than the average wholesale price. Production of tomato in a polyhouse was not feasible, as the breakeven price was more than the average market price and all the project appraisal parameters indicated that it was not feasible. Only at about 48% premium price over the prevailing market price or reduction of cost of polyhouse structure by 60% from 400 to 160 Rs./m², one could make the tomato production viable in a polyhouse.

## 4.2.4 CRITIQUE ON REVIEW OF LITERATURE

From the forgoing review, it is observed that considerable research work was carried out on irrigation scheduling on various vegetable crops under polyhouse. Enough literature was not available on water requirement, irrigation, and fertigation scheduling of capsicum crop and systematic studies. The area under polyhouse and drip irrigation are increasing day by day and capsicum appears to be well adopted to this method but further research is needed to determine optimal crop water requirement, right irrigation, and fertigation scheduling to maximize yield and quality produce under polyhouse. Field studies inside polyhouse are required to evaluate the response of drip irrigation and fertigation scheduling on biometric parameters of yield and fruit yield of capsicum and its overall economics. Not enough research has been conducted in this direction. Efforts have been made in this chapter to consider these aspects of drip irrigation and fertigation scheduling to investigate the response on yield and overall economics of capsicum production under polyhouse.

## 4.3 MATERIALS AND METHODS

### 4.3.1 MICRO ENVIRONMENTAL PARAMETERS

Experiments were conducted in two different types of NVPHs to study effects of tensiometer-based drip irrigation scheduling of capsicum under polyhouse.

To study various micro environmental parameters, observations taken inside the polyhouses were the temperature, RH, and solar intensity. The temperature and RH was recorded every day at 9:00, 12:00, and 16:00 while the maximum and minimum temperature was recorded every morning at 9:00 AM. The NVPH was equipped with: thermometers on the right and left side from entrance to the last corner to record the temperature, hygrometer for recording the RH, and maximum and minimum temperature-recording thermometer at the center. The solar intensity was also recorded at 9:00, 12:00, and 16:00 inside the polyhouse, every day by using digital lux meter (Model: LX-101)

### 4.3.2  ESTIMATION OF WATER REQUIREMENT OF CAPSICUM UNDER NVPH

#### 4.3.2.1  ESTIMATION OF REFERENCE EVAPOTRANSPIRATION

The study was undertaken with the objective to estimate the crop water requirement of capsicum in double span naturally ventilated polyhouse (DS NVPH) and walking tunnel type naturally ventilated polyhouse (WT NVPH).

The reference evapotranspiration $(ET_0)$ was estimated using the FAO-PM (1998) method from the available data of temperature, RH and wind speed, and sunshine hours.

FAO-PM method is recommended as the sole $ET_0$ method for determining reference ET in a wide range of location and climates and had provision for application in situation where limited data were available.

A particular form of PM equation developed by Smith et al. [64] known as FAO-PM equation had been used to estimate daily $ET_0$. The final form of the FAO-PM equation was as given below,

$$ET_0 = \frac{0.408\Delta\,(R_n - G) + \gamma\dfrac{900}{T + 278}u_z(e_s - e_a)}{\Delta + \gamma\,(1 + 0.34u_z)}, \qquad (4.6)$$

where $ET_0$ = reference evapotranspiration (mm/day); $\Delta$ = slope of saturation vapor pressure curve (kPa·C$^{-1}$); $R_n$ = net radiation (MJ·m$^{-2}$·day$^{-1}$); $G$ = soil heat flux density (MJ·m$^{-2}$·day$^{-1}$); $\gamma$ = psychometric constant (kPa·C$^{-1}$); T = mean daily air temperature (°C); $e_s$ = saturation vapor pressure (kPa); $e_a$ = actual vapor pressure (kPa); and $u_2$ = average daily wind speed at 2-m height (m/s).

Calculation of $ET_0$ using eq. (4.6) on daily basis requires meteorological data consisting of maximum and minimum daily air temperatures ($T_{max}$ and $T_{min}$), mean daily actual vapor pressure ($e_a$) derived from dew point temperature or RH ($Rh_1$ and $Rh_2$) data, daily average of 24 h of wind speed measured at 2-m height ($u_2$), net radiation ($R_n$) measured or computed from solar and long-wave radiation or the actual duration of sunshine hours ($n$). The extraterrestrial radiation and day light hours ($n$) for specific day of the month were also computed. As the magnitude of soil heat flux ($G$) beneath the reference grass surface was relatively small, it was ignored for daily time stake.

### 4.3.2.2   ESTIMATION OF CROP EVAPOTRANSPIRATION (ET_c)

Crop ET under standard conditions is defined as the ET from disease free, well fertilized crops, grown in large fields, under optimum field conditions and achieving full production under the given climatic conditions.

The crop ET differs distinctly from the reference evapotranspiration ($ET_0$) as the ground cover, canopy properties and aerodynamic resistance of the crop are different from the grass. The effects of characteristics, that distinguish field crops from grass, are integrated into the crop coefficient ($K_c$). In the crop coefficient approach, crop ET is calculated by multiplying $ET_0$ by $K_c$ (FAO–56). Based on this, crop ET under the polyhouses was calculated by multiplying $ET_0$ by $K_c$.

$$ET_{crop} = K_c * ET_0. \qquad (4.7)$$

In eq (4.7): $ET_{crop}$ = crop evapotranspiration (mm/day), $K_c$ = crop coefficient (dimensionless), $ET_0$ = reference crop evapotranspiration (mm/day). The crop coefficients, $K_c$, were used based on the FAO-56 curve methods. The crop coefficient values depend on the type of crop and its growing stage, growing season, and the prevailing weather conditions. The shape of the curve represents the changes in the vegetation and ground cover during plant development and maturation that affect the ratio of $ET_c$ to $ET_0$. The $K_c$ values for the capsicum under the study were used daily from Fig. 4.1.

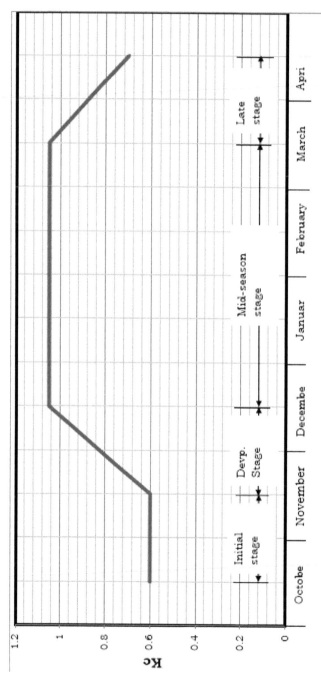

**FIGURE 4.1**   Estimation of crop coefficient for capsicum.

## 4.3.3 DRIP IRRIGATION AND FERTIGATION SCHEDULING OF CAPSICUM UNDER POLYHOUSE

The experiments were conducted during October 6, 2010 to April 30, 2011 in silty clay loam soil at the Department Irrigation and Drainage Engineering, College of Technology, G. B. Pant University of Agriculture and Technology, Pantnagar. The site is located in the Terai region of Uttarakhand state (India) at the foot hills of the Shivalik range of Himalayas and lies at 29.5°N latitude, 79.3°E longitudes and at an altitude of 243.83 m above mean sea level.

Geographically, Pantnagar, comes under the humid subtropical zone with average annual rainfall of 1400 mm with the monsoon season of 4 months. The summer is too dry and hot, the winter is too cold, and rainy season with a heavy rainfall. The dry season is from November to May and wet or monsoon season comes after mid-June till mid-October. The monsoon normally starts in the second week of June and continues with appreciate strength up to the August, with its peak in July. The RH remains high at about 90% up to February, after which it decreases up to 55% in April and that remains more or less constant up to the onset of monsoon. The mean minimum monthly temperature swings from 5°C to 25°C while the mean maximum temperature varies 20–40°C. The weather parameters recorded at meteorological observatory, crop research centre, Pantnagar, located at about 0.5 km away from the experimental site were utilized for the estimation of the water requirement of the crop under study.

The top 30 cm of the soil profile is Beni-silty clay loam with sand (18%), silt (50%) and clay (32%). The average ground water table of the experimental site is 60–100 cm.

A gravity-fed drip system consisted of a water tank, wire mesh screen filter, PVC control valve, main line, submain, laterals, take-off pipe connectors, end cap, etc. and for irrigation scheduling with the help of 12 tensiometers (at 6 in-depth each) that were installed in both the polyhouses. For recording the micro environmental data inside the polyhouses, thermometers, maximum–minimum thermometers, hygrometers, lux meter, etc. were used.

### 4.3.3.1 LAYOUT

Two polyhouses were selected for the experiment. Both are NV but their shapes and area are different. The covering material for both the polyhouses was 200 micron UV-stabilized polyethylene.

*DS NVPH* was of 20 m × 10 m having a single door and equipped with side wall role up curtains on both side walls to cover the provided net for natural environmental ventilation and the main purpose of net at the side walls to protect the plants from insect, pests, and to enter the cool air inside the polyhouse (Figs. 4.2 and 4.3). The cladding material was made of UV-stabilized polyethylene sheet of 200 microns (800 gauges). It

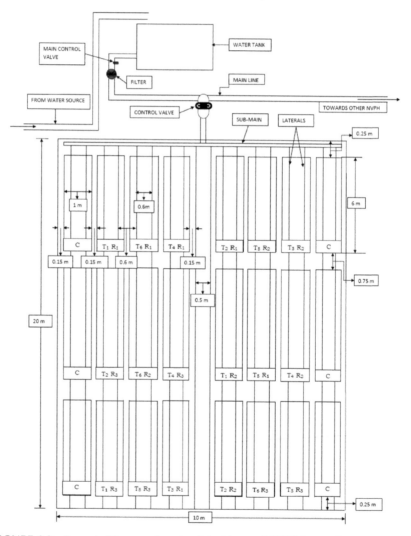

**FIGURE 4.2**   Layout of the experiment and double span NVPH for capsicum crop under drip irrigation and irrigation scheduling based on readings of tensiometers.

had two span gable roofs and central height was about 5 m from ground surface. The orientation of polyhouse was in E–W direction. In natural ventilation, air becomes less dense as it gets heated and rises up. Thus warm air moves out and allows the cool air to flow into the polyhouse through the net at the side wall curtains.

**(A)**

**(B)**

**FIGURE 4.3**   Views of double span naturally ventilated polyhouse (DV NVPH): A (top): outside; and B (bottom): inside.

There was also provided a light shade net to reduce high light intensity in the summer season. And in the winter season side wall curtains fully covered the net to increase the inside temperature. The NVPH was equipped with eight thermometers, four on the right and four on the left to record the temperature inside, besides hygrometer for recording the RH and maximum and minimum temperature recording thermometer at the center. And irrigation scheduling based on plant–soil water tension that was measured by six stationary vacuum gauge tensiometers (installed, three at right and three at left).

Size of *walking tunnel-type naturally ventilated polyhouse* (WT NVPH) was 20 m × 5 m (Figs. 4.4 and 4.5). It was equipped with side wall roll up curtains on both side walls as provided on DS NVPH and having a single door at the middle with curved roof, whose radius of curvature was about 1.8 m. The cladding material was made of UV-stabilized polyethylene film of 200 microns (800 gauges). Central height of curved roof was about 2.5 m from ground surface.

The orientation of WT polyhouse is in the E–W direction. The rolling up of side wall curtains allows the flow of cool air inside the polyhouse and to move out the warm air from polyhouse. And in the summer season, light shed nets were used to cover the curved roof to reduce the high light intensity due to the sun to plants. In the winter season, rolling up of side wall curtains was able to fully cover the net. The WT NVPH was equipped with six thermometers (three on right and three on left) to record the temperature of polyhouse, besides hygrometer for recording the RH and maximum–minimum temperature-recording thermometer at the center. Irrigation scheduling was based on the readings of six stationary vacuum gauge tensiometers (three at right and three at left), that were installed at 20 cm depth.

### 4.3.3.2  EXPERIMENTAL DETAILS FOR DS NVPH AND WT NVPH

The study included three irrigation levels: $I_1 = 20–30$ kPa; $I_2 = 30–50$ kPa; and $I_3 = 50–70$ kPa in both the polyhouses. Among the readings of tensiometer, minimum tension point indicated end of irrigation and maximum reading indicated the start of irrigation. This was able to maintain the irrigation level at a specified tension in all three irrigation treatments. These three irrigation levels were combined with two fertigation levels. By the combination of three irrigation levels and two fertigation levels, six main treatments and a controlled treatment were used for the experimentation in both the polyhouses.

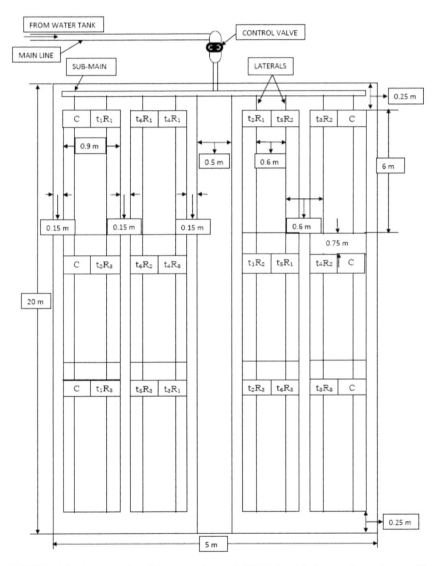

**FIGURE 4.4** Layout of walking tunnel type NVPH for drip-irrigated capsicum with irrigation scheduling based on readings of tensiometers.

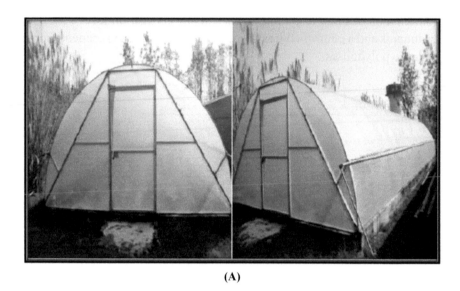

**(A)**

**(B)**

**FIGURE 4.5** Views of walking tunnel-type naturally ventilated polyhouse (WT NVPH): A (top): outside; and B (bottom): inside.

### 4.3.3.2.1  Treatment Details of DS NVPH

The total area 200 m² of the field inside DS NVPH was divided into 24 plots and each plot size was 6 m × 1 m. Each plot consisted of two rows of plants and each row consisted of 18 plants. The placements of the treatments on the experimental plots with three replications were allotted based on randomized block design methods. 18 of 24 plots were main treatment plots and remaining 6 plots were under controlled treatment. Tensiometer were installed in between the row of each middle plot. Irrigation treatments begun 2 weeks after transplanting to allow the plant to establish the root system and to bring the plant–soil moisture tension up to desired level (i.e., maximum tension point). Irrigation started at maximum tension point and continued up to minimum tension point by daily tensiometer readings. Amount of water calculated by how much of water applied through drip tape considering the minimum tension point and maximum tension point. The fertilizer application based on soil analysis result the recommended dose of fertilizer 300:60:400 kg/ha of NPK was used. The 50% dose of nitrogen, complete dose of $P_2O_5$ and 50% dose of $K_2O$ were applied as basal dose before transplanting and remaining applied in split doses at interval of 7 days from 2 weeks after transplanting and up to the end of the crop. The details of irrigation and fertigation scheduling are shown in Tables 4.1 and 4.2. The weeding, cultivation, pest and disease control measures, and other cultural operations were carried out as necessary.

**TABLE 4.1**  Details of the Treatments in DS NVPH and WT NVPH.

| Treatment, I.D. | | Replications | Irrigation level (kPa) | Fertigation level (%) |
|---|---|---|---|---|
| DS NVPH | WT NVPH | | | |
| $I_1F_1(T_1)$ | $I_1F_1(t_1)$ | $R_1$ | 20–30 | 75 |
| | | $R_2$ | | |
| | | $R_3$ | | |
| $I_1F_2(T_2)$ | $I_1F_2(t_2)$ | $R_1$ | 20–30 | 100 |
| | | $R_2$ | | |
| | | $R_3$ | | |
| $I_2F_1(T_3)$ | $I_2F_1(t_3)$ | $R_1$ | 30–50 | 75 |
| | | $R_2$ | | |
| | | $R_3$ | | |
| $I_2F_2(T_4)$ | $I_2F_2(t_4)$ | $R_1$ | 30–50 | 100 |
| | | $R_2$ | | |
| | | $R_3$ | | |

**TABLE 4.1**   *(Continued)*

| Treatment, I.D. | | Replications | Irrigation level (kPa) | Fertigation level (%) |
|---|---|---|---|---|
| **DS NVPH** | **WT NVPH** | | | |
| | | $R_1$ | | |
| $I_3F_1(T_5)$ | $I_3F_1(t_5)$ | $R_2$ | 50–70 | 75 |
| | | $R_3$ | | |
| | | $R_1$ | | |
| $I_3F_2(T_6)$ | $I_3F_2(t_6)$ | $R_2$ | 50–70 | 100 |
| | | $R_3$ | | |
| $C(T_7)$ | $C(t_7)$ | – | – | – |

F1 (75%) → N:$P_2O_5$:$K_2O$:225:45:300 kg/ha.
F2 (100%) → N:$P_2O_5$:$K_2O$:300:60:400 kg/ha.
F = fertigation; I = irrigation; C = control.

### 4.3.3.2.2   Treatment Details of WT NVPH

The total area of 100 m² inside WT NVPH was divided into 12 plots (6 at the left and 6 at the right) with a buffer of 0.5 m space and in the middle (Fig. 4.3). Each bed size was 6 m × 0.9 m consisting of two rows of plants and each row contained 18 plants. Tensiometer was installed in each treatment row except the control treatment row. Layout design, irrigation, and fertigation (Tables 4.1 and 4.3) scheduling procedures, other measurements, and cultural operations were exactly same as in DS NVPH.

#### 4.3.3.2.2.1   Agronomical Details

The test crop was bell peppers (*Capsicum annum* L. cv. Swarna F1 hybrid). Before sowing the seeds, the soil was prepared by mixing from 67% of field soil, 22% vermicompost and 11% of paddy husk. Then the mixture was filled into potting plugs of 15 trays with each tray containing (14 × 7) 98 holes with a total of 1470 holes. Each hole could hold 40–50 g of soil. The seed was treated for 24 hours with plant biocontrol agent-3 (*Tricoderma harzianum* + *Pseudomonas fluorescens*) at 10g/1g seed. The seeds were sown in another NVPH of experimental site on 10 September,

**TABLE 4.2** Details of Fertigation Scheduling in DS NVVH.

| Week after transplanting | $F_1$ | | $F_2$ | | Remaining % applied after basal dose | Dosages |
|---|---|---|---|---|---|---|
| | N | K | N | K | | |
| 3–6 | 0.703 kg | — | 0.796 kg | — | 50% N | 3.26 kg Urea (0-0-46) in 4 splits |
| 7–10 | 0.281 kg | 0.375 kg | 0.318 kg | 0.425 kg | 20% N, 20% K | 1.74 kg $KNO_3$ (13-0-46) and 0.8 kg Urea in 4 splits |
| 11–13 | — | 0.281 kg | — | 0.318 kg | 20% K | 1.2 kg NPK (0-0-50) in 3 splits |
| 14 | 0.07kg | 0.090 kg | 0.080 kg | 0.110 kg | 5% NPK | 1.1 kg NPK (19-19-19) |
| 15–30 | 0.350 kg | 1.125 kg | 0.40 kg | 1.275 kg | 25% N, 60% K | a. 2.4 kg NPK (0-0-50) in 8 splits; b. 2.6 kg $KNO_3$, 0.9 kg Urea in 8 splits; c. alternate week (K, NK) |

**TABLE 4.3**  Details of Fertigation Scheduling in WT NVVH.

| Week after transplanting | $F_1$ | | $F_2$ | | Remaining % applied after basal dose | Dosages |
|---|---|---|---|---|---|---|
| | N | K | N | K | | |
| 3–6 | 0.350 kg | – | 0.398 kg | – | 50% N | 1.63 kg Urea (0-0-46) in 4 splits |
| 7–10 | 0.140 kg | 0.188 kg | 0.159 kg | 0.213 kg | 20% N, 20% K | 0.87 kg $KNO_3$ (13-0-46) & 0.41 kg Urea in 4 splits |
| 11–13 | – | 0.140 kg | – | 0.159 kg | 20% K | 0.6 kg NPK (0-0-50) in 3 splits |
| 14 | 0.035 kg | 0.045 kg | 0.04 kg | 0.055 kg | 5% NPK | 0.52 kg NPK (19-19-19) |
| 15–30 | 0.175 kg | 0.563 kg | 0.20 kg | 0.638 kg | 25% N, 60% K | a. 1.2 kg NPK (0-0-50) in 8 splits. b. 1.3 kg $KNO_3$, 0.45 kg Urea in 8 splits. c. alternate week (K, NK) |

2010. After sowing, the soil in each tray hole was treated with suspension of bio-agent at 10 kg/L of water. Then the trays were watered regularly till the transplantation. The germination rates of the seeds were 75%. The seedlings were ready for transplanting after 25 days from the date of sowing and were transplanted on October 6, 2010 in DS NVPH and October 08, 2010 in WT NVPH, respectively. Drenching of each plant with suspension of bio-agent at 10 g/L of water was done immediately after transplanting. Gap filling was done on 10th day after transplanting by putting new seedlings at open spaces to maintain the optimum plant population.

The plants in both polyhouses were supervised regularly. Depending on the observed symptoms for fungus, bacteria or any insects, the protection, or control measures were taken.

### 4.3.3.2.2.2 *Details of Biometric Observations*

Five observational plants, which are healthy, medium, and average in height, were randomly selected from each replication and were tied aluminum tag for easy identification. Biometric observations included plant height, number of branches, average diameter, average length, and weight of fruits were taken.

The height of plant in each treatment and in each replication was made at 30-days interval starting after 30 days of transplanting. The plant height was measured from the bed to the highest leaf of the individual plant.

The number of branches from the observational plant was counted, whereas the undeveloped branches were not considered in counting.

These observations were taken during picking of fresh sweet pepper fruits, and observations of each and every fruit were taken from observational plants only.

The volume displacement method was used to obtain the volume of fruit. The density of the fruit (g/cm$^3$) was calculated by following equation:

$$\text{Density of fruit} = [(\text{fruit weight})/(\text{fruit volume})] \qquad (4.8)$$

The number of fruits was recorded from the tagged plants. The total number of fruits picked from five tagged plants from each replication gave the average number of fruits/plant in each replication of each treatment.

The matured fruits were weighed and total weight divided by the number of fruits gave average weight of fruit. From observational plant, weights of each fruit were taken from each treatment and in each replication.

### 4.3.3.3  CROP YIELD

The first picking of the capsicum fruit in DS NVPH was done on December 27, 2010, 82 DAT and the second picking was on January 24, 2011, 110 DAT while the last (7th) picking was done on May 2, 2011, 208 DAT. And in WT NVPH, first picking was done on January 7, 2011, 91 DAT and the second picking was on February 2, 2011, 117 DAT while the last (7th) picking was done on May 6, 2011, 210 DAT. The diameter, length, weight and volume of each fruit were taken at each picking. The total yield of each treatment was then calculated in kg/m$^2$.

#### 4.3.3.3.1  Water Use Efficiency

The WUE for each treatment was calculated by dividing the total yield obtained in each treatment to the total depth of water application during its growing period.

$$\text{WUE, kg/m}^3 = [\text{yield, kg/m}^2]/[\text{total irrigation depth, m}] \qquad (4.9)$$

### 4.3.3.4  ECONOMIC FEASIBILITY ANALYSIS OF CAPSICUM PRODUCTION UNDER NATURALLY VENTILATED POLYHOUSES

The cost of the polyhouse depends on the type of crop, its spacing and the system to be used. The BCR was calculated for capsicum under tensiometer-based irrigation at three irrigation levels with two fertigation levels in both polyhouses. The life period of the polyhouse was taken as 24 years. The initial cost of the polyhouse includes the cost of all components of polyhouse including the structural, cladding material, pumping unit, filter, and distribution networks of irrigation system. The depreciation cost was assumed to be 10% of the total cost and a rate of interest was taken as 13.5%.

The total cost was then worked out for 6 months. The cost of cultivation of capsicum was worked out for 6 months, which included expenses incurred on land preparation, cost of seed, nursery management, application of fertilizers, plant protection measures, and labor costs.

For evaluation the BCR, the yield and selling price of produce was used. The selling price rate was taken from the nearby local market.

## 4.4  RESULTS AND DISCUSSION

The study was undertaken with the objective to investigate the effects of tensiometer-based drip irrigation scheduling on capsicum (*Capsicum annum* L.) under polyhouse. This section presents the results obtained from the study undertaken during October 6, 2010 to April 30, 2011 and has been categorized into three main sections. In the first section, various microclimatic parameters, the estimation of reference evapotranspiration ($ET_0$) and the water requirement under DS-NVPH and WT-NVPH have been discussed. The second section is divided into two subsections, in the first subsection the amount of water applied at different irrigation levels based on tensiometer reading in DS-NVPH and WT-NVPH has been elucidated. In the second subsection, effects of irrigation and fertigation scheduling on the growth, yield and yield attributes, the WUE, and the water productivity for capsicum are presented. While the last section deals with the economic feasibility of production of colored capsicum under tensiometer-based drip irrigation and fertigation scheduling in naturally ventilated polyhouses with variable market price.

### 4.4.1  MICRO ENVIRONMENTAL PARAMETERS AND WATER REQUIREMENT OF CAPSICUM UNDER NATURALLY VENTILATED POLYHOUSES

#### 4.4.1.1  VARIATION OF VARIOUS CLIMATIC PARAMETERS UNDER DS-NVPH AND WT-NVPH

The observations were carried out in the naturally ventilated polyhouses to study the variation in the microclimatic parameters such as: temperature ($T_{max}$, $T_{min}$, and T), relative humidity (Rh), and solar intensity.

For open-field conditions, daily maximum and minimum temperatures, actual sunshine hours, wind velocity, and relative humidity were collected from the University Observatory located at Crop Research Center, Pantnagar. The daily temperature (°C), relative humidity (%), and solar intensity (Lux) at various times in DS-NVPH and WT-NVPH are presented.

The daily variation of maximum and minimum temperatures in DS-NVPH, WT-NVPH, and open-field conditions is shown in Figures 4.6–4.8. It can be inferred that maximum temperature trend is shown by WT-NVPH followed by DS-NVPH, and then finally open field occupies the last position. The maximum temperature recorded in WT-NVPH was 45.2°C on November 14, 2010 and 43°C in DS-NVPH on November 17, 2010 and 40°C on April 30, 2011 in open field. The trend in open field was lower than both polyhouses up to March 15, 2011 and from March 16, 2011 onwards, it was quite similar to both polyhouses. On March 16, 2011, temperature was 31.6°C in DS-NVPH and 31°C in WT-NVPH and open field. The minimum temperature is almost highest in WT-NVPH and quite similar trend is followed by DS-NVPH until December 4, 2010 and from this date onwards fluctuation occurred in both trends. The temperature trend for open field was always found lower except at some points. The minimum temperature in open field was 2.8°C on January 19, 2011, it was 6.7°C on December 27, 2011 in DS-NVPH and 10.3°C on January 13, 2011 in WT-NVPH.

The deviations in maximum and minimum daily temperatures of DS-NVPH and WT-NVPH over open-field condition are shown in Figure 4.7. In WT-NVPH, the deviation in maximum daily temperature over open-field condition is higher up to 21°C on December 3, 2010 while the minimum deviation in maximum daily temperature over open field was −5°C on April 30, 2011. The maximum temperature during this time in open field was higher than maximum temperature of both the polyhouses. In DS-NVPH, the maximum daily temperature deviation was 28.5°C on December 22, 2010 while the minimum deviation was −4.4°C on April 30, 2011.

In case of deviation in minimum daily temperature in DS-NVPH, except at some points which was maximum up to 22°C on December 7, 2010 and minimum up to −3°C on March 19, 2011 over open field. The temperature trend deviations in minimum daily temperature in WT-NVPH started from less than 5°C, and increased up to 21.6°C on December

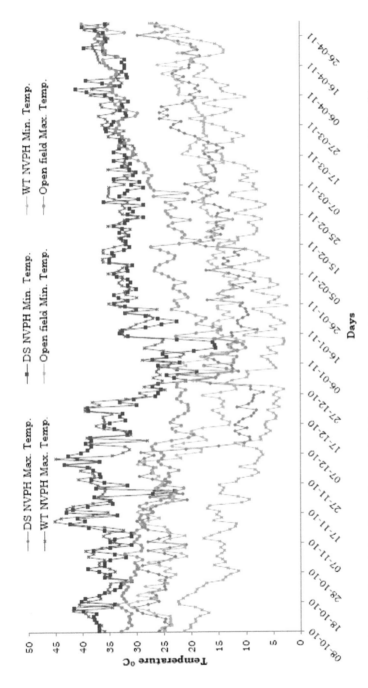

**FIGURE 4.6** Daily variations in maximum and minimum temperatures in DS-NVPH, WT-NVPH and open-field Condition.

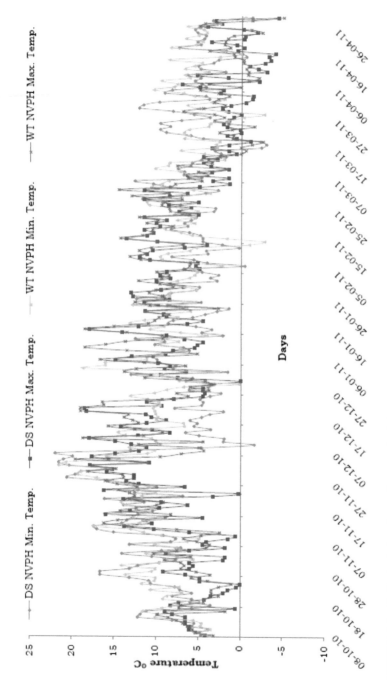

**FIGURE 4.7** Daily deviations in maximum and minimum temperatures in DS-NVPH and WT-NVPH over open field.

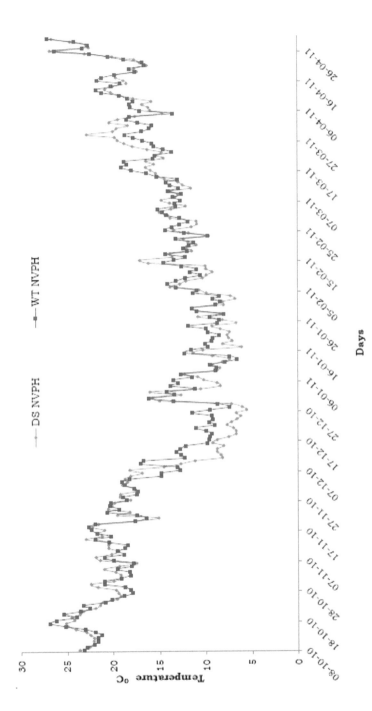

**FIGURE 4.8**   Daily variations in temperature at 9:00 a.m. in DS-NVPH and WT-NVPH.

3, 2010, and decreased up to −2.8°C on February 15, 2011. During the growing period, it was observed that the temperature for optimal growth rate for capsicum crop was 16°C–25°C. A gradual rise in temperature from 25°C and above causes dropping of flowers, buds, fruit and leaf burn while it creates favorable conditions especially for fruit borers that mainly attack immature leaves first and then start boring of fruits that is mostly observed in naturally ventilated polyhouses. The temperature below 15°C resulted in lower flower production, fruiting, and stunt growth of the crop. The temperature requirements for bell pepper are higher than those for tomatoes. Fruit-set does not occur below 16°C or above 32°C and maximum fruit-set takes place in between 16°C and 21°C. Bud and flower abscission is a major problem during summer.

The daily variation of temperature inside DS-NVPH and WT-NVPH recorded at 9:00 a.m., 12, and 4:00 p.m. is shown from Figures 4.8–4.10. The variation in temperature at 9:00 a.m. varied from maximum of 26.85°C (April 26, 2011) to minimum of 5.5°C (December 27, 2010) for DS-NVPH compared with 27.1°C (April 17, 2010) to 6.6°C (January 13, 2011) for WT-NVPH. The general trends of the daily temperature at 9:00 a.m. in DS-NVPH and WT-NVPH were initially in the increasing order up to October 16, 2010 and later decreased from October 17, 2010 to December 27, 2010, but higher than 15°C and then increased from March 16, 2011. The average temperature trend of WT-NVPH is almost higher than that of DS-NVPH and from April 18, 2011 up to April 30, both trends followed closely the same path (Fig. 4.8).

Figure 4.9 reveals that the daily variation in temperature at 12 noon was maximum on November 14, 2010 (37.1°C) and the minimum was on January 8, 2011 (16.8°C), for WT-NVPH and it was almost higher than the temperature of DS-NVPH throughout the growing period. For DS-NVPH, maximum temperature was 37.7°C on April 8, 2011 and the minimum was 12.75°C on January 8, 2011. Both trends followed almost same pattern which means variation in both temperature trends inside DS-NVPH and WT-NVPH at 12 noon were same throughout the growing period.

The daily variation in temperature at 16:00 was maximum of 35.27°C on October 17, 2010 and the minimum of 15.57°C on January 13, 2011 in WT-NVPH, while in DS-NVPH the maximum was 30.57°C on October 16, 2010 and the minimum was 10.27°C on January 13, 2011 (Fig. 4.10). As it is observed in this figure at 16:00, the temperature

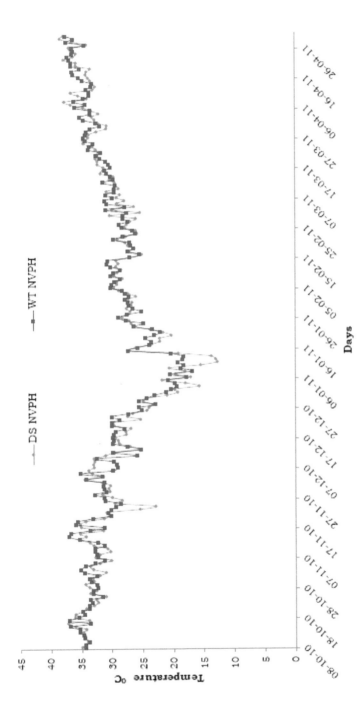

**FIGURE 4.9** Daily variations in temperature at 12:00 noon in DS-NVPH and WT-NVPH.

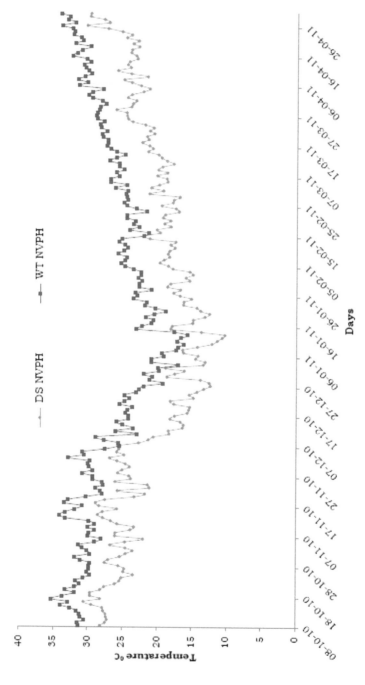

**FIGURE 4.10**   Daily variations in temperature at 4:00 p.m. in DS-NVPH and WT-NVPH.

in WT-NVPH is higher than DS-NVPH except on December 7, 2010 temperature was almost the same, DS-NVPH (25.57°C) and WT-NVPH (25.43°C). Higher temperature under WT-NVPH might be due to its shape and volume.

The daily variation of relative humidity in DS-NVPH and WT-NVPH recorded at 9:00 a.m., 12:00 noon, and 4:00 p.m. is shown from Figures 4.11–4.13. Figure 4.11 shows that the relative humidity at 9:00 a.m. was higher in WT-NVPH and followed by DS-NVPH. The daily relative humidity % trend on October 8, 2010 was 67% for WT-NVPH and 54% for DS-NVPH. The trend was almost higher in WT-NVPH except for some days between the start of January including few days of first to second week. More than 75% relative humidity inside W-NVPH was found on November 17, 2010 at the start of winter season which was rapidly decreased up to 61% on November 29, 2010 and at constant rate of 65%.

From December 3–6, 2010, similarly, in DS-NVPH for same duration, it was 61%. From that point onwards both the trends hold the same path. The maximum relative humidity % in WT-NVPH was 81% on February 16, 2011 and minimum was 37% on April 10, 2011, while for DS-NVPH it was maximum 65% on December last and January last and minimum was 35% on April 17, 2011. Due to deposition of dew on the roof of the polyhouses, the humidity % was increased above 80% and it also resulted in the increase of moisture content of soil as well. Due to this, there was attack of fungus *Sclerotinia*, stem rot, and blight which decayed any part of the stem of the plant and blocked the transport functions and ultimately caused wilting of the plants. This was observed on few plants in both polyhouses. As compared with WT-NVPH, the relative humidity in the DS-NVPH was relatively better for optimal growth of the crop because of its shape, size, and volume.

In Figure 4.12, it can be observed that the relative humidity at 12:00 noon was higher in WT-NVPH followed by DS-NVPH. Deviation between both trends was 5–10% of relative humidity throughout the season. The trend started on October 8, 2010 from 42% relative humidity for WT-NVPH and 35% for DS-NVPH. The maximum relative humidity at 12:00 noon in WT-NVPH was 49% on December 19–21, 2010 and minimum was 26% on April 17, 2011. In DS-NVPH, the maximum relative humidity was 42% on January 6, 2011 and minimum was 23% on April 30, 2011. There was a constant humidity observed on the trend for few days in both the polyhouses.

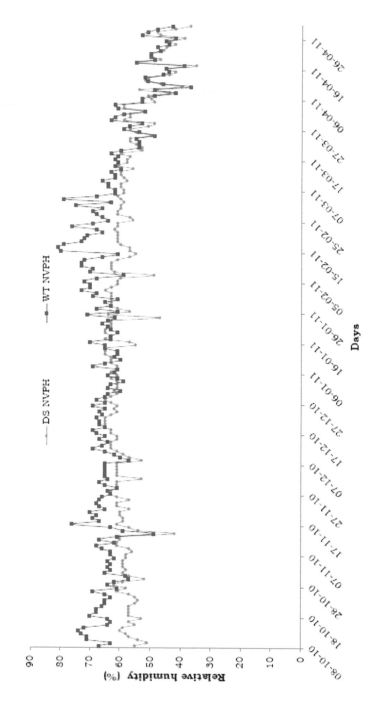

**FIGURE 4.11**   Daily variations in relative humidity at 9:00 a.m. in DS-NVPH and WT-NVPH.

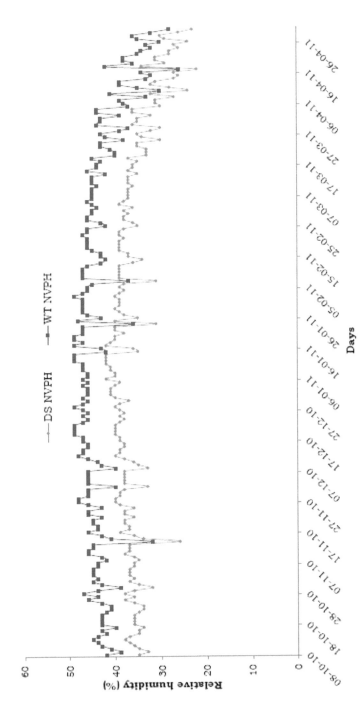

**FIGURE 4.12** Daily variation in relative humidity at 12:00 noon in DS-NVPH and WT-NVPH.

Figure 4.13 indicates that the relative humidity at 16:00 was higher in WT-NVPH and followed by DS-NVPH which was almost lower throughout the season. The trend started from 61% relative humidity for WT-NVPH and 50% on October 8, 2010. The constant trend was 56% of relative humidity on December 3–7, 2010 for WT-NVPH. Similarly, it was 52% on December 3–7, 2010 for DS-NVPH. The maximum relative humidity in WT-NVPH was 79% on February 16, 2011 and minimum was 29% on April 10, 2011. The maximum relative humidity in DS-NVPH was 63% on January 12–13, 2011 whereas the minimum was 30% on April 17, 2011. The variation in relative humidity was due to the outer environmental conditions such as: cloudy atmosphere, rainy and winter season, and other conditions that had high moisture content.

The daily variations of solar intensity recorded at 9:00 a.m., 12:00 noon, and 16:00 in DS-NVPH and WT-NVPH are shown in Figures 4.14–4.16.

Figure 4.14 reveals that the solar intensity at 9:00 a.m. was higher in WT-NVPH followed by DS-NVPH. The trend for WT-NVPH started from 41.0 kLux and 25.0 kLux for DS-NVPH on October 8, 2010. Variation in solar intensity was same in both trends: initial gradual variation in WT-NVPH from starting to October 18, 2010 to reach up to 51.0 kLux and rapid variation was found in DS-NVPH from starting to October 18, 2010, then it increased up to 45.20 kLux. The maximum solar intensity for WT-NVPH was 66.50 kLux on November 20, 2010 and minimum was 2.20 kLux on March 3, 2011. For DS-NVPH, maximum was 52.30 kLux on November 21, 2010 and minimum was 1.70 kLux on December 31, 2010. Due to more humid, cloudy, and rainy conditions, there was a decrease in values of solar intensities. The higher solar intensity in WT-NVPH might be due to its lower control height as compared with DS-NVPH.

Figure 4.15 shows that the solar intensity at 12:00 noon was higher in WT-NVPH followed by DS-NVPH. The trend was started from 68.0 kLux for WT-NVPH on October 8, 2010 and 48.0 kLux for DS-NVPH on October 8, 2010. The maximum solar intensity for WT-NVPH was 89.90 kLux on April 5, 2011 and the minimum was 2.80 kLux on December 31, 2010. In DS-NVPH, the minimum was 2.20 kLux on December 30, 2010 and maximum was 68.0 kLux on February 18, 2011. Figure 4.16 shows that the solar intensity at 4:00 p.m. was higher in WT-NVPH followed by DS-NVPH. The starting trend for WT-NVPH was 18.50 kLux on October 8, 2010 and maximum rising up to 51.20 kLux on February 18, 2011, and minimum reaching up to 1.20 kLux. For DS-NVPH, the trend started from

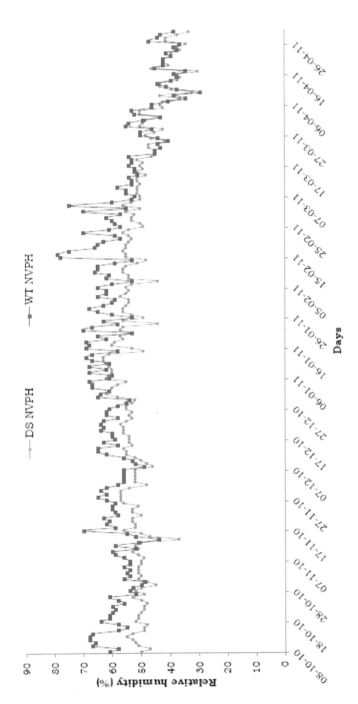

**FIGURE 4.13** Daily variations in relative humidity at 4:00 p.m. in DS-NVPH and WT-NVPH.

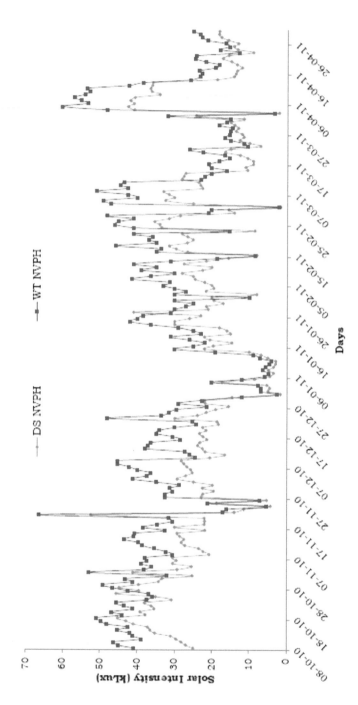

**FIGURE 4.14** Daily variations in solar intensity at 9:00 a.m. in DS-NVPH and WT-NVPH.

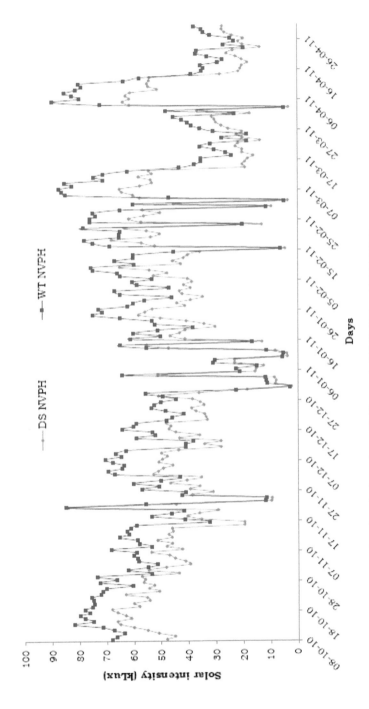

**FIGURE 4.15** Daily variations in solar intensity at 12:00 noon in DS-NVPH and WT-NVPH.

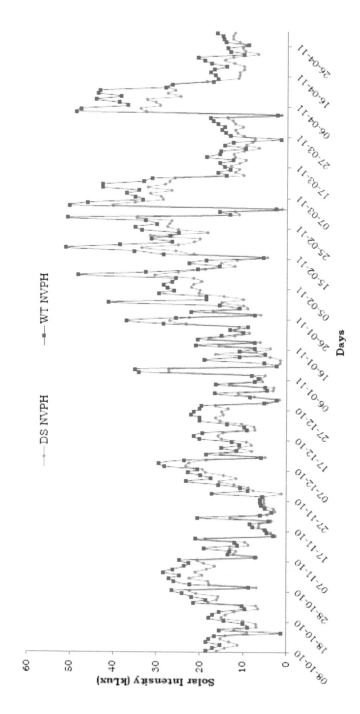

**FIGURE 4.16**   Daily variation in solar intensity at 4:00 p.m. in DS-NVPH and WT-NVPH.

16.0 kLux on October 8, 2010, maximum up to 40.0 kLux on March 4, 2011, and just before this day minimum solar intensity was 1.2 kLux on March 3, 2011.

The solar intensity at 12:00 noon was generally higher than the observations at 9:00 a.m. and 16:00. The solar intensity in both NVPH reached to its maximum on third week of February, and for controlling the high solar intensity it was supplied with shed net as ceiling.

In general during January 1–January 31, the solar intensity was very low and due to this crop growth was stunt, flowering rate and fruit production were reduced because solar intensity to the level of 25.0 kLux or more is required for the optimum growth of the plant. The solar intensity under the polyhouse is of utmost importance for the growth of plant as well as to estimate crop water requirement. The variation in solar intensity at various times of the day was mainly due to the daylight hours.

## 4.4.2 REFERENCE EVAPOTRANSPIRATION AND WATER REQUIREMENT OF CAPSICUM IN DS-NVPH AND WT-NVPH

The daily reference evapotranspiration ($ET_0$) for DS-NVPH and WT-NVPH was estimated using PM model and values are presented in Figure 4.17.

The pattern of reference evapotranspiration in Figure 4.17 gradually decreased from October 17, 2010 to January 13, 2011 and thereafter it had an increasing trend. The reference evapotranspiration was higher in WT-NVPH as compared with DS-NVPH throughout the period of estimation. The trend was almost same for both NVPH with the deviation of 0.35–0.55 mm/day. The higher temperature and solar intensity could be the main factors for high reference evapotranspiration under WT-NVPH as compared with DS-NVPH.

The maximum $ET_0$ for WT-NVPH was 4.73 mm/day on April 20, 2011 and minimum was 0.93 mm/day on January 3, 2011. For DS-NVPH, maximum was 4.67 mm/day on April 20, 2011 and minimum was 0.84 mm/day two times on January 3 and 11, 2011. The average reference evapotranspiration in DS-NVPH was 2.36 mm/day compared with 2.49 mm/day in WT-NVPH.

The daily crop water requirement of capsicum under drip irrigation in DS-NVPH and WT-NVPH was estimated using FAO-56 approach and values are presented in Figure 4.18. The pattern of crop water requirement in DS- and WT-NVPH was in the decreasing order from October 17

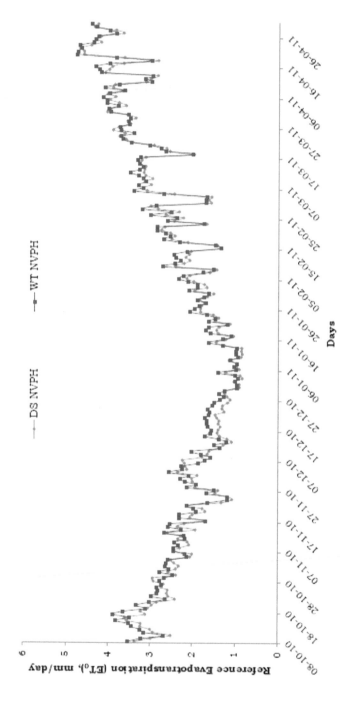

**FIGURE 4.17**   Daily variations of reference evapotranspiration ($ET_0$) in DS-NVPH and WT-NVPH.

to November 5, 2010 and fluctuated up to the end of December, 2010 and thereafter it was continually increasing up to April 4, 2011 and then decreasing trend was observed for both NVPH. The crop water requirement in WT-NVPH was higher as compared with DS-NVPH. The average water requirement during the study period was 0.34 L/day/plant in DS-NVPH and 0.36 L/day/plant in WT-NVPH. The maximum water requirement in WT-NVPH was 0.69 L/day/plant on April 4 and 5, 2011 and minimum was 0.16 L/day/plant on December 31, 2010, January 1, 3, 8, 10, and 13, 2011. The water requirement in WT-NVPH was generally higher, due to its shape and lower volume than DS-NVPH. The estimated crop evapotranspiration was 4.6% higher under WT-NVPH compared with DS-NVPH.

### 4.4.3 WATER REQUIREMENT OF CAPSICUM UNDER NATURALLY VENTILATED POLYHOUSES

#### 4.4.1 2 THREE IRRIGATION DEPTHS BASED ON TENSIOMETERS

The three levels of water depth based on readings of tensiometers were: $I_1$ (20–30 kPa), $I_2$ (30–50 kPa), $I_3$ (50–70 kPa) under tensiometers and a control plot. The performance of capsicum under DS-NVPH and WT-NVPH was evaluated and the data are plotted in Figures 4.19 and 4.20. The cumulative amounts of water application are shown as a function of day of growing season for all four irrigation treatments. The steeper slopes of the curves under control treatment, showed time periods of high water demand, similar trend was also obtained under the tensiometer-based irrigation treatments during the same period of observation.

Figures 4.19 and 4.20 show that initially up to 15 days after transplanting (DAT), the irrigation was kept constant for development of root and establishment of the capsicum plant in all treatments. At this stage in each of the three irrigation levels and control, cumulative irrigation depth was 4.95 L/plant. Cumulative amount of water depth in the control plot was higher in both polyhouses compared to tensiometer-based three irrigation levels.

In Figure 4.19, it can be observed that under DS-NVPH the amount of water applied in control plot was 29.68%, 59.8%, and 77.23% higher than $I_1$, $I_2$, and $I_3$ levels of irrigation, respectively. As shown in Figure 4.20 for WT-NVPH, it was 22.77%, 63.8%, and 80.51% higher in the respective treatments. Table 4.4 shows that the total amount of water applied in

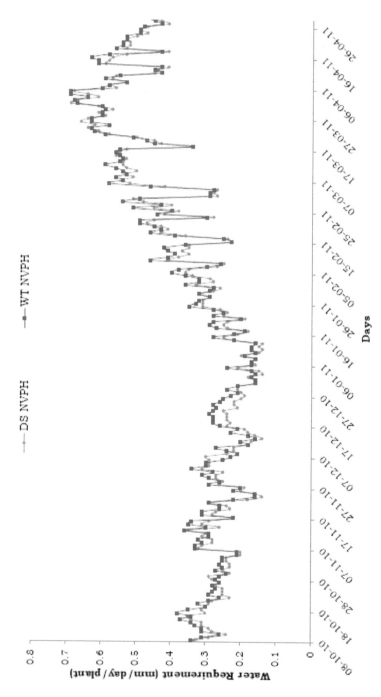

**FIGURE 4.18**   Daily variations of estimated water requirements in DS-NVPH and WT-NVPH.

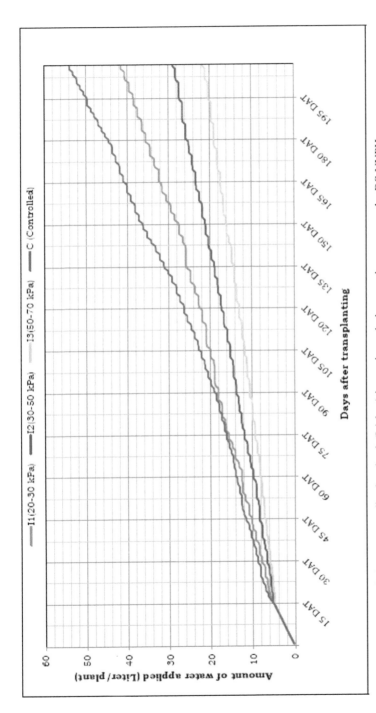

**FIGURE 4.19**   Cumulative water application depth (L/plant) in capsicum during growing season under DS-NVPH.

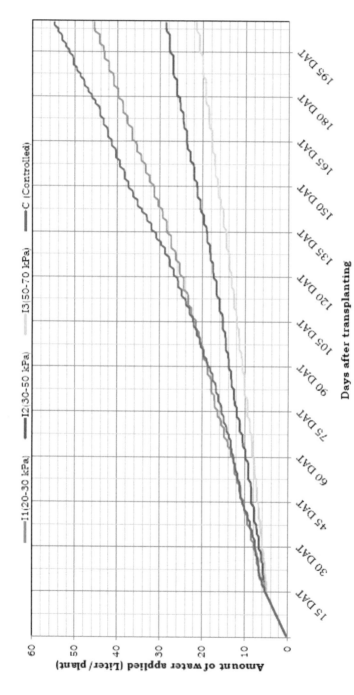

**FIGURE 4.20**  Cumulative water application (L/plant) in capsicum during growing season under WT-NVPH.

tensiometer-based different irrigation levels $I_1$, $I_2$, and $I_3$ and control plot was 41.51, 29.01, 21.77, and 53.83 L/plant, respectively, in DS-NVPH as compared with 45.49, 28.64, 21.52, and 54.94 L/plant in WT-NVPH, respectively. The data pertaining to the water saving and relative water use due to tensiometer-based irrigation scheduling over control are given in Table 4.4. The water saving (%) and relative water use in case of $I_3$, $I_2$, and $I_3$ levels of irrigation over control were 17–23%, 46–48%, and 60–61%, respectively, and the relative water use was 0.17–0.23, 0.46–0.48, and 0.60–0.61 in the respective irrigation levels. Water saving due to tensiometer-based irrigation scheduling was higher because the water application based on soil moisture tension involves least water loss and chance of over-irrigation was minimized. In control treatment, the water application was based on conventional irrigation practices at 25% soil moisture depletion, therefore there are chances of over-irrigation.

**TABLE 4.4** Total Water Application, Corresponding Water Saving, and Relative Water Use as Compared with Control.

| Treatments | Water applied (L/plant) | | Water saving (%) over control | | Relative water use | |
|---|---|---|---|---|---|---|
| | DS-NVPH | WT-NVPH | DS-NVPH | WT-NVPH | DS-NVPH | WT-NVPH |
| $I_1$ | 41.51 | 45.49 | 23 | 17 | 0.23 | 0.17 |
| $I_2$ | 29.01 | 28.64 | 46 | 48 | 0.46 | 0.48 |
| $I_3$ | 21.77 | 21.52 | 60 | 61 | 0.60 | 0.61 |
| C | 53.83 | 54.94 | – | – | – | – |

## 4.4.4 RESPONSE OF CAPSICUM TO DRIP IRRIGATION AND FERTIGATION SCHEDULING UNDER DS-NATURALLY VENTILATED POLYHOUSES: BIOMETRIC PARAMETERS, YIELD, AND YIELD ATTRIBUTES

Table 4.5 indicates that effects of different treatments on plant height of capsicum were significant at 30 and 60 DAT and nonsignificant at 90 DAT. At 30 DAT, the plant height was maximum in the treatment $T_6$ which was 38.29% higher than control $T_7$. At 60 DAT, it was maximum in the treatment $T_5$ which was 32.55% higher than control $T_7$.

Table 4.5 shows that the effect of different treatments on stem diameter was found to be nonsignificant at 30 DAT and significant at 60 and 90 DAT observations. Stem diameter of capsicum plant at 60 DAT was observed maximum in the treatment $T_6$ which was 31.46% higher than control $T_7$ and at 90 DAT it was observed maximum in the treatment $T_4$ which was 36.11% higher.

As it was observed from Table 4.5, in the different treatments, the critical difference was found to be significant on crop canopy at 30 and 60 DAT and nonsignificant at 90 DAT. At 30 DAT, the crop canopy was maximum 45.12% higher in the treatment $T_4$ and at 60 DAT it was maximum 29.99% higher in the treatment $T_5$ as compared with the control $T_7$.

### 4.4.4.1 NUMBER OF PRIMARY AND SECONDARY BRANCHES

Table 4.5 shows that the effects of different treatments on number of primary branches were nonsignificant in all observations at 30, 60, and 90 DAT. Also Table 4.5 reveals that the effects of the treatments on number of secondary branches were nonsignificant at 30 DAT and significant at 60 and 90 DAT. Number of secondary branches at 60 DAT was maximum in the treatment $T_1$, 23.97% higher than the control $T_7$. At 90 DAT, it was also maximum and 33.72% higher in the treatment $T_6$ than $T_7$. Figure 4.21 shows the effects of different treatments on vegetative growth of capsicum at 75 DAT in DS-NVPH.

**TABLE 4.5** Effects of Various Treatments on Capsicum Plant Height, Stem Diameter, Canopy Width, and Number of Branches at 30, 60, and 90 DAT in DS-NVPH.

| Treatment | 30 DAT | % Increase over control | 60 DAT | % Increase over control | 90 DAT | % Increase over control |
|---|---|---|---|---|---|---|
| | | | Plant height (cm) | | | |
| $T_1$ | 42.47 | 29.17 | 72.20 | 23.44 | 83.93 | 19.52 |
| $T_2$ | 42.80 | 30.17 | 68.20 | 16.60 | 80.30 | 14.35 |
| $T_3$ | 43.67 | 32.82 | 75.67 | 29.37 | 87.30 | 24.32 |
| $T_4$ | 44.40 | 35.04 | 71.53 | 22.29 | 93.33 | 32.91 |
| $T_5$ | 44.20 | 34.43 | 77.53 | 32.55 | 89.80 | 27.88 |
| $T_6$ | 45.47 | 38.29 | 73.53 | 25.71 | 91.97 | 30.97 |

**TABLE 4.5** *(Continued)*

| Treatment | 30 DAT | % Increase over control | 60 DAT | % Increase over control | 90 DAT | % Increase over control |
|---|---|---|---|---|---|---|
| $T_7$ | 32.88 | – | 58.49 | – | 70.22 | – |
| CD (*P* < 0.05) | **6.44** | | **7.05** | | NS | |
| SEM ± | 2.09 | | 2.29 | | 4.73 | |
| CV | 8.56 | | 5.58 | | 9.61 | |
| | | | **Stem diameter (cm)** | | | |
| $T_1$ | 0.46 | 21.05 | 1.09 | 22.47 | 1.27 | 17.59 |
| $T_2$ | 0.46 | 21.05 | 1.07 | 20.22 | 1.17 | 8.33 |
| $T_3$ | 0.50 | 31.58 | 1.07 | 20.22 | 1.22 | 12.96 |
| $T_4$ | 0.51 | 34.21 | 1.16 | 30.34 | 1.47 | 36.11 |
| $T_5$ | 0.48 | 26.32 | 1.13 | 26.97 | 1.29 | 19.44 |
| $T_6$ | 0.49 | 28.95 | 1.17 | 31.46 | 1.40 | 29.63 |
| $T_7$ | 0.38 | – | 0.89 | – | 1.08 | – |
| CD (*P* < 0.05) | NS | | 0.086 | | **0.20** | |
| SEM ± | 0.039 | | 0.028 | | 0.066 | |
| CV | 11.76 | | 4.45 | | 9.01 | |
| | | | **Crop canopy width** | | | |
| $T_1$ | 42.67 | 16.68 | 59.93 | 10.45 | 70.90 | 8.41 |
| $T_2$ | 40.50 | 10.75 | 67.47 | 24.35 | 80.89 | 23.69 |
| $T_3$ | 44.73 | 22.31 | 63.07 | 16.24 | 79.57 | 21.67 |
| $T_4$ | 53.07 | 45.12 | 66.26 | 22.12 | 78.76 | 20.43 |
| $T_5$ | 43.87 | 19.96 | 70.53 | 29.99 | 81.23 | 24.20 |
| $T_6$ | 49.47 | 35.27 | 63.13 | 16.35 | 79.27 | 21.21 |
| $T_7$ | 36.57 | – | 54.26 | – | 65.40 | – |
| CD (*P* < 0.05) | **8.37** | | **9.03** | | NS | |
| SEM ± | 2.72 | | 2.93 | | 3.60 | |
| CV | 10.59 | | 7.99 | | 8.15 | |
| | | | **Number of primary branches** | | | |
| $T_1$ | 2.90 | 30.63 | 3.36 | 20.00 | 3.99 | 20.91 |
| $T_2$ | 2.33 | 4.95 | 2.93 | 4.64 | 3.38 | 2.42 |
| $T_3$ | 2.67 | 20.27 | 3.16 | 12.86 | 4.32 | 30.91 |
| $T_4$ | 2.57 | 15.77 | 3.55 | 26.79 | 3.97 | 20.30 |
| $T_5$ | 2.69 | 21.17 | 3.40 | 21.43 | 3.87 | 17.27 |

**TABLE 4.5**  *(Continued)*

| Treatment | 30 DAT | % Increase over control | 60 DAT | % Increase over control | 90 DAT | % Increase over control |
|---|---|---|---|---|---|---|
| $T_6$ | 2.81 | 26.58 | 3.77 | 34.64 | 4.20 | 27.27 |
| $T_7$ | 2.22 | – | 2.80 | – | 3.30 | – |
| CD ($P < 0.05$) | NS | | NS | | NS | |
| SEM ± | 0.18 | | 0.26 | | 0.29 | |
| CV | 12.52 | | 13.82 | | 13.06 | |
| **Number of secondary branches** | | | | | | |
| $T_1$ | 5.30 | 23.83 | 6.00 | 23.97 | 7.18 | 19.87 |
| $T_2$ | 4.76 | 11.21 | 5.45 | 12.60 | 6.45 | 7.68 |
| $T_3$ | 5.35 | 25.00 | 5.97 | 23.35 | 7.34 | 22.54 |
| $T_4$ | 4.97 | 16.12 | 5.86 | 21.07 | 7.84 | 30.88 |
| $T_5$ | 5.13 | 19.86 | 5.63 | 16.32 | 6.37 | 6.34 |
| $T_6$ | 5.40 | 26.17 | 5.96 | 23.14 | 8.01 | 33.72 |
| $T_7$ | 4.28 | – | 4.84 | – | 5.99 | – |
| CD ($P < 0.05$) | NS | | 0.69 | | 0.78 | |
| SEM ± | 0.31 | | 0.22 | | 0.25 | |
| CV | 10.84 | | 6.86 | | 6.22 | |

**FIGURE 4.21**   Effects of different treatments on vegetative growth of capsicum at 75 DAT in DS-NVPH.

### 4.4.4.2   FRUIT LENGTH/DIAMETER/DENSITY

Table 4.6 shows that the average fruit length of capsicum in different treatments was significant. The average length of fruit in the treatments $T_2$ and $T_3$ was increased by 26.80% as compared with the control $T_7$. It was also observed that the average diameter of fruit in different treatments was significant. The average diameter of fruit was increased maximum by 27.06% in the treatments $T_1$ and $T_6$ than control $T_7$. As seen from Table 4.6, the effects of different treatments on the average fruit density were nonsignificant.

**TABLE 4.6**   Effects of Various Treatments on Fruit Parameters of Capsicum Fruits in DS-NVPH.

| Treatment | Fruit length | % increase over control | Fruit diameter | % increase over control | Fruit density | % increase over control |
|---|---|---|---|---|---|---|
| $T_1$ | 10.74 | 26.80 | 9.29 | 25.71 | 0.51 | -10.53 |
| $T_2$ | 10.74 | 26.80 | 9.19 | 24.36 | 0.54 | -5.26 |
| $T_3$ | 10.42 | 23.02 | 9.02 | 22.06 | 0.56 | -1.75 |
| $T_4$ | 10.54 | 24.44 | 9.17 | 24.09 | 0.55 | -3.51 |
| $T_5$ | 10.66 | 25.86 | 9.39 | 27.06 | 0.55 | -3.51 |
| $T_6$ | 10.74 | 26.80 | 9.29 | 25.71 | 0.51 | -10.53 |
| $T_7$ | 8.47 | – | 7.39 | – | 0.57 | – |
| CD ($P < 0.05$) | 0.52 | | 0.57 | | NS | |
| SEM ± | 0.17 | | 0.19 | | 0.034 | |
| CV | 2.84 | | 3.60 | | 10.88 | |

### 4.4.4.3   AVERAGE FRUIT WEIGHT/NUMBER OF FRUITS PER PLANT/YIELD PER PLANT

In double span NVPH, the average weight of fruit was 251.67 g in tensiometer-based irrigation with fertigation treatment and in control it was 210.00 g. Table 4.7 shows that the effects of treatments on the average fruit weight were significant. The average weight of fruit was increased maximum in the treatments $T_1$ and $T_2$ by 25.40% as compared with control $T_7$. Table 4.7 shows that the effects of the treatments on the number of fruits per plant were nonsignificant. The mean yield per plant was 2.03 kg/plant in the tensiometer-based irrigation with fertigation treatment and in control it was 1.42 kg/plant. The effects of various treatments on the average yield per plant were nonsignificant. View of yield and vegetative growth of capsicum in DS-NVPH is shown in Figure 4.22.

**TABLE 4.7**   Effects of Various Treatments on Yield Parameters of Capsicum Crop in DS-NVPH.

| Treatment | Fruit weight (g) | % Increase over control | No. of fruits per plant | % Increase over control | Yield per plant (kg) | % Increase over control |
|---|---|---|---|---|---|---|
| $T_1$ | 263.33 | 25.40 | 7.65 | 13.00 | 2.00 | 40.85 |
| $T_2$ | 263.33 | 25.40 | 7.82 | 15.51 | 2.06 | 45.07 |
| $T_3$ | 240.00 | 14.29 | 7.50 | 10.78 | 1.82 | 28.17 |
| $T_4$ | 240.00 | 14.29 | 8.57 | 26.59 | 2.05 | 44.37 |
| $T_5$ | 250.00 | 19.05 | 8.73 | 28.95 | 2.19 | 54.23 |
| $T_6$ | 253.33 | 20.63 | 8.27 | 22.16 | 2.09 | 47.18 |
| $T_7$ | 210.00 | – | 6.77 | – | 1.42 | – |
| CD ($P < 0.05$) | 26.47 | | NS | | NS | |
| SEM ± | 8.60 | | 0.70 | | 0.18 | |
| CV | 6.05 | | 15.34 | | 15.08 | |

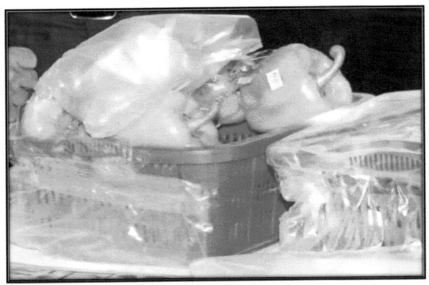

**FIGURE 4.22** View of yield and vegetative growth of capsicum in DS-NVPH.

## 4.4.4.4 YIELD PER SQUARE METER/WATER USE EFFICIENCY/ WATER PRODUCTIVITY

In the double span naturally ventilated polyhouse (DS-NVPH) from six tensiometer-based irrigation with fertigation treatments, the average yield per square meter was 12.20 kg/m$^2$ and maximum production was 13.14 kg/m$^2$ in the treatment $T_5$ or $I_3F_1$ while the minimum was 10.92 kg/m$^2$ in the treatment $T_3$ or $I_2F_1$. While in control, the yield was 8.52 kg/m$^2$ lowest compared with all the treatments under drip irrigation. Table 4.8 shows that the effects of the different treatments on yield per meter square were nonsignificant. The effects of the various treatments were significant on WUE. The WUE was maximum in the treatment $T_5$ by 281.31% higher as compared with the control $T_7$. Table 4.8 reveals that the effects of the various treatments on water productivity (L/kg) were significant. The water productivity in treatment $T_6$ was decreased maximum by −72.43% as compared with control $T_7$. The amount of water required to produce 1 kg of capsicum was maximum for the control treatment (37.98% L/kg) and minimum in treatment $T_5$ (10.15% L/kg). In general, water productivity under drip irrigated treatment was higher (45.34–73.28%) than surface irrigated treatment.

**TABLE 4.8**  Effects of Various Treatments on Capsicum Yield Per Meter Square, Water Use Efficiency (WUE), and Water Productivity in DS-NVPH.

| Treatment | Yield (kg/m2) | % Increase over control | WUE (kg/m3) | % Increase over control | Water productivity (L/kg) | % Increase over control |
|---|---|---|---|---|---|---|
| $T_1$ | 12.00 | 40.85 | 48.18 | 82.64 | 20.76 | -45.34 |
| $T_2$ | 12.34 | 44.84 | 49.55 | 87.83 | 20.33 | -46.47 |
| $T_3$ | 10.92 | 28.17 | 62.74 | 137.83 | 17.07 | -55.06 |
| $T_4$ | 12.28 | 44.13 | 70.55 | 167.44 | 14.40 | -62.09 |
| $T_5$ | 13.14 | 54.23 | 100.59 | 281.31 | 10.15 | -73.28 |
| $T_6$ | 12.54 | 47.18 | 96.00 | 263.91 | 10.47 | -72.43 |
| $T_7$ | 8.52 | – | 26.38 | – | 37.98 | – |
| CD ($P < 0.05$) | NS | | 20.47 | | 4.56 | |
| SEM ± | 1.06 | | 6.64 | | 1.48 | |
| CV | 15.80 | | 17.74 | | 13.68 | |

### 4.4.5  RESPONSE OF CAPSICUM CROP TO DRIP IRRIGATION AND FERTIGATION SCHEDULING UNDER WT-NATURALLY VENTILATED POLYHOUSE: BIOMETRIC PARAMETERS, YIELD, AND YIELD ATTRIBUTES

It is seen from Table 4.9 that the effects of different treatments on capsicum plant height were significant at 30, 60, and 90 DAT. At 30 DAT, the plant height was maximum in treatment $t_1$, which was 41.58% higher. At 60 DAT, maximum 21.20% and at 90 DAT maximum 22.09% were higher in treatment $t_1$ as compared with the control treatment $t_7$. Table 4.9 shows that the effects of different treatments on stem diameter were significant at 30, 60, and 90 DAT. At 30 DAT, the maximum stem diameter was observed in the treatment $t_6$ which was 30.56% higher than control $t_7$, at 60 DAT it was maximum 27.59% higher in the treatment $t_5$ and at 90 DAT it was maximum 30.28% higher in the treatment $t_1$. Table 4.9 shows that cop canopy width in different treatments was nonsignificant at 30 DAT on crop canopy and significant at 60 and 90 DAT. At 60 DAT, the maximum crop canopy was observed in treatment $t_1$ which was 23.32% higher and at 90 DAT it was maximum 19.99% higher in treatment $t_6$ as compared with control $t_7$. Table 4.9 shows that the effect of different treatments on number of primary branches was nonsignificant in all observations at 30, 60, and 90 DAT. Also the effect of the different treatments on number of secondary branches was nonsignificant at 30 and 60 DAT and significant at 90 DAT. Number of secondary branches at 90 DAT was maximum 25.57% higher in the treatment $t_2$ than control $t_7$. View of vegetative growth of capsicum at 75 DAT in WT-NVPH is shown in Figure 4.23.

**TABLE 4.9** Effects of Irrigation Scheduling Based on Tensiometer and Fertigation on Height, Stem Diameter, Crop Canopy, and Number of Branches of Capsicum Plant under WT-NVPH at 30, 60, and 90 DAT.

| Treatment | 30 DAT | % Increase over control | 60 DAT | % Increase over control | 90 DAT | % Increase over control |
|---|---|---|---|---|---|---|
| | | | Plant height (cm) | | | |
| $t_1$ | 45.46 | 41.58 | 72.88 | 21.20 | 89.25 | 22.09 |
| $t_2$ | 38.94 | 21.27 | 68.5 | 13.92 | 82.65 | 13.06 |
| $t_3$ | 40.28 | 25.44 | 68.90 | 14.59 | 82.30 | 12.59 |
| $t_4$ | 44.31 | 37.99 | 71.65 | 19.16 | 86.92 | 18.91 |
| $t_5$ | 42.96 | 33.79 | 71.06 | 18.18 | 86.70 | 18.60 |
| $t_6$ | 44.88 | 39.77 | 71.43 | 18.79 | 88.16 | 20.60 |
| $t_7$ | 32.11 | – | 60.13 | – | 73.10 | – |
| CD ($P < 0.05$) | 0.88 | | 4.56 | | 8.88 | |
| SEM ± | 0.29 | | 1.48 | | 2.88 | |
| CV | 1.20 | | 3.71 | | 5.93 | |
| | | | Stem diameter (cm) | | | |
| $t_1$ | 0.43 | 19.44 | 1.10 | 26.44 | 1.42 | 30.28 |
| $t_2$ | 0.44 | 22.22 | 1.07 | 22.99 | 1.34 | 22.94 |
| $t_3$ | 0.43 | 19.44 | 1.10 | 26.44 | 1.41 | 29.36 |
| $t_4$ | 0.42 | 16.67 | 1.05 | 20.69 | 1.31 | 20.18 |
| $t_5$ | 0.45 | 25.00 | 1.11 | 27.59 | 1.33 | 22.02 |
| $t_6$ | 0.47 | 30.56 | 1.08 | 24.14 | 1.35 | 23.85 |
| $t_7$ | 0.36 | – | 0.89 | – | 0.87 | – |
| CD ($P < 0.05$) | 0.043 | | 0.086 | | 0.086 | |
| SEM ± | 0.014 | | 0.028 | | 0.028 | |
| CV | 5.56 | | 4.45 | | 4.62 | |

**TABLE 4.9**   *(Continued)*

| Treatment | 30 DAT | % Increase over control | 60 DAT | % Increase over control | 90 DAT | % Increase over control |
|---|---|---|---|---|---|---|
| | | | **Crop canopy width (cm)** | | | |
| $t_1$ | 46.67 | 17.82 | 67.96 | 23.32 | 76.37 | 18.15 |
| $t_2$ | 45.32 | 14.42 | 67.30 | 22.12 | 77.24 | 19.49 |
| $t_3$ | 47.58 | 20.12 | 64.95 | 17.86 | 75.41 | 16.66 |
| $t_4$ | 45.85 | 15.75 | 60.77 | 10.27 | 72.01 | 11.40 |
| $t_5$ | 48.57 | 22.62 | 63.64 | 15.48 | 77.65 | 20.13 |
| $t_6$ | 45.60 | 15.12 | 64.41 | 16.88 | 77.56 | 19.99 |
| $t_7$ | 39.61 | – | 55.11 | – | 64.64 | – |
| CD ($P < 0.05$) | NS | | **2.15** | | **5.17** | |
| SEM ± | 2.56 | | 0.69 | | 1.67 | |
| CV | 9.71 | | 1.90 | | 3.90 | |
| | | | **Number of primary branches** | | | |
| $t_1$ | 5.37 | 18.81 | 7.08 | 31.60 | 7.59 | 15.53 |
| $t_2$ | 5.06 | 11.95 | 5.94 | 10.41 | 8.25 | 25.57 |
| $t_3$ | 5.48 | 21.24 | 6.59 | 22.49 | 8.08 | 22.98 |
| $t_4$ | 5.26 | 16.37 | 5.96 | 10.78 | 7.60 | 15.68 |
| $t_5$ | 5.11 | 13.05 | 5.91 | 9.85 | 7.33 | 11.57 |
| $t_6$ | 5.59 | 23.67 | 6.54 | 21.56 | 7.50 | 14.16 |
| $t_7$ | 2.30 | – | 2.73 | – | 3.63 | – |
| CD ($P < 0.05$) | NS | | NS | | NS | |
| SEM ± | 0.22 | | 0.17 | | 0.23 | |
| CV | 14.60 | | 9.53 | | 10.23 | |
| | | | **Number of secondary branches** | | | |
| $t_1$ | 5.37 | 18.81 | 7.08 | 31.60 | 7.59 | 15.53 |
| $t_2$ | 5.06 | 11.95 | 5.94 | 10.41 | 8.25 | 25.57 |
| $t_3$ | 5.48 | 21.24 | 6.59 | 22.49 | 8.08 | 22.98 |
| $t_4$ | 5.26 | 16.37 | 5.96 | 10.78 | 7.60 | 15.68 |
| $t_5$ | 5.11 | 13.05 | 5.91 | 9.85 | 7.33 | 11.57 |
| $t_6$ | 5.59 | 23.67 | 6.54 | 21.56 | 7.50 | 14.16 |
| $t_7$ | 4.52 | – | 5.38 | – | 6.57 | – |
| CD ($P < 0.05$) | NS | | NS | | 0.97 | |
| SEM ± | 0.25 | | 0.35 | | 0.31 | |
| CV | 8.4 | | 9.6 | | 7.2 | |

**FIGURE 4.23** View of vegetative growth of capsicum at 75 DAT in WT-NVPH.

As seen in Table 4.10, the average fruit length in different treatments was significant. The average length of fruit was increased and was maximum by 28.35% in treatment $t_3$ as compared with control $t_7$. Table 4.10 reveals the average diameter of fruit in different treatments and the values were significant. The average diameter of fruit was increased and was maximum by 32.63% in the treatment $t_6$ compared with $t_7$. As seen from Table 4.10, the different treatments affected the average fruit density nonsignificantly.

In WT-NVPH, the average weight of fruit was 236.7 g in tensiometer-based irrigation with fertigation treatment compared with 200 g in control. Table 4.11 shows that the effects of the treatments on the average fruit weight were significant. The average weight of fruit was increased and was maximum in the treatment $t_2$ by 33.34% as compared with control $t_7$. Table 4.11 also shows that the effect of the treatments on the number of fruits per plant was significant. The number of fruits per plant was increased and was maximum by 34.50% in the treatment $t_4$ as compared with the control $t_7$. In WT-NVPH, the average yield per plant was 2.12 kg/plant in the tensiometer-based irrigation with fertigation treatment compared with 1.44 kg/plant in the control treatment. From Table 4.11 one can conclude that the effect of various treatments on average yield per plant was significant. The yield per plant was increased and was maximum by 107.64% in the treatment $t_6$ as compared with the control $t_7$.

**TABLE 4.10**  Effects of Various Treatments on Fruit Parameters of Capsicum Fruits in WT-NVPH.

| Treatment | Fruit length | % Increase over control | Fruit diameter | % Increase over control | Fruit density | % Increase over control |
|---|---|---|---|---|---|---|
| $t_1$ | 10.86 | 18.43 | 8.90 | 25.18 | 0.52 | −7.14 |
| $t_2$ | 10.34 | 12.76 | 8.75 | 23.07 | 0.69 | 23.21 |
| $t_3$ | 11.77 | 28.35 | 8.78 | 23.49 | 0.55 | −1.79 |
| $t_4$ | 10.47 | 14.18 | 8.32 | 17.02 | 0.48 | −14.29 |
| $t_5$ | 10.83 | 18.10 | 9.15 | 28.69 | 0.51 | −8.93 |
| $t_6$ | 10.51 | 14.61 | 9.43 | 32.63 | 0.49 | −12.50 |
| $t_7$ | 9.17 | – | 7.11 | – | 0.56 | – |
| CD ($P < 0.05$) | **0.86** | | **0.92** | | **NS** | |
| SEM ± | 0.28 | | 0.29 | | 0.043 | |
| CV | 4.58 | | 5.96 | | 13.85 | |

**TABLE 4.11**  Effects of Various Treatments on Yield Parameters of Capsicum Crop in WT-NVPH.

| Treatment | Fruit weight (g) | % Increase over control | No. of fruits per plant | % Increase over control | Yield per plant (kg) | % Increase over control |
|---|---|---|---|---|---|---|
| $t_1$ | 266.67 | 33.34 | 7.81 | 8.93 | 2.07 | 43.75 |
| $t_2$ | 253.33 | 26.67 | 6.85 | −4.46 | 1.81 | 25.69 |
| $t_3$ | 210.00 | 5.00 | 9.63 | 34.31 | 2.02 | 40.28 |
| $t_4$ | 217.00 | 8.50 | 9.02 | 25.80 | 1.96 | 36.11 |
| $t_5$ | 245.00 | 22.50 | 8.53 | 18.97 | 2.99 | 107.64 |
| $t_6$ | 266.67 | 33.34 | 7.81 | 8.93 | 2.07 | 43.75 |
| $t_7$ | 200.00 | – | 7.17 | – | 1.44 | – |
| CD ($P < 0.05$) | **25.37** | | **1.59** | | **0.45** | |
| SEM ± | 8.24 | | 0.52 | | 0.14 | |
| CV | 6.16 | | 10.96 | | 12.38 | |

Table 4.12 shows that the different treatments affected the yield per meter square of capsicum nonsignificantly. In the WT-NVPH for six tensiometer-based irrigation schedulings with fertigation, the maximum production was 13.49 kg/m$^2$ in the treatment $t_4$ or $I_2F_2$ while the minimum was 11.94 kg/m$^2$ in the treatment $t_5$ or $I_3F_1$. In the control plot, yield was lowest (9.56 kg/m$^2$) as compared to all drip irrigated treatments. As seen in Table 4.12, the effect of the various treatments on WUE was significant. The WUE was increased and was maximum in the treatment $t_5$ by 248.05% as compared with the control $t_7$. Table 4.12 reveals that the effect of the various treatments on water productivity was found to be significant. The water productivity was decreased maximum in the treatment $t_5$ by −71.45% as compared with control $t_7$. The amount of water applied to produce 1 kg of capsicum was maximum (38.53 L/kg) for the surface irrigated control treatment whereas the amount of water required to produce 1 kg of capsicum ranged between 11.00 and 24.60 L/kg in the drip irrigated treatments. View of yield and vegetative growth of capsicum in WT-NVPH are shown in Figure 4.24.

**TABLE 4.12**   Effects of Various Treatments on Capsicum Yield Per Meter Square, Water Use Efficiency (WUE), and Water Productivity in WT-NVPH.

| Treatment | Yield (kg/m2) | % Increase over control | WUE (kg/m3) | % Increase over control | Water productivity (L/kg) | % Increase over control |
|---|---|---|---|---|---|---|
| $t_1$ | 13.81 | 44.46 | 45.53 | 74.31 | 22.01 | −42.88 |
| $t_2$ | 11.67 | 22.07 | 81.37 | 211.52 | 12.94 | −66.42 |
| $t_3$ | 13.49 | 41.11 | 70.66 | 170.52 | 14.20 | −63.15 |
| $t_4$ | 13.04 | 36.40 | 90.91 | 248.05 | 11.00 | −71.45 |
| $t_5$ | 11.94 | 24.90 | 83.19 | 218.49 | 12.08 | −68.65 |
| $t_6$ | 13.81 | 44.46 | 45.53 | 74.31 | 22.01 | −42.88 |
| $t_7$ | 9.56 | – | 26.12 | – | 38.53 | – |
| CD ($P < 0.05$) | NS | | **17.51** | | **4.12** | |
| SEM ± | 0.89 | | 5.68 | | 1.34 | |
| CV | 12.96 | | 15.71 | | 11.98 | |

**FIGURE 4.24**  View of yield and vegetative growth of capsicum in walking tunnel-type naturally ventilated polyhouse.

## 4.4.6 ECONOMIC RETURNS OF PRODUCTION OF COLORED CAPSICUM UNDER TENSIOMETER-BASED IRRIGATION IN NATURALLY VENTILATED POLYHOUSES

The economic analysis for the production of colored capsicum was worked out under tensiometer-based drip irrigation scheduling under polyhouses.

The life period of the polyhouse was taken as 24 years. The initial cost of the polyhouse included the cost of all components of polyhouse including structural material, cladding material, shade net, pumping unit, filter, and distribution networks of irrigation system. The depreciation cost was assumed as 10% of the total cost whereas the rate of interest was taken as 13.5%. The cost of cultivation of capsicum included the expenses incurred on land preparation, seed, nursery management, vermin-compost, soil solarization, fertilizers, plant protection measures, power required for lifting water. During the growing period, the market price of colored capsicum varied between 20 and 120 Rs. and it was used to evaluate the BCR and net seasonal income (NSI). The BCR at different market prices for the production of colored capsicum in 20 m × 10 m DS-NVPH and 20 m × 5 m WT-NVPH is shown in Tables 4.13 and 4.14.

The trends of the results for the different treatments in DS-NVPH and WT-NVPH at different market prices are shown in Figures 4.25 and 4.26.

Tables 4.13 and 4.14 and Figures 4.25 and 4.26 indicate that at the selling price of 20 Rs./kg, the BCR for tensiometer-based irrigation scheduling with fertigation treatments in DS-NVPH and WT-NVPH ranged between 0.96 and 1.55 out of which more than one was economically feasible. In all treatments of DS-NVPH and WT-NVPH at 40 Rs./kg of colored capsicum, BCR ranged between 1.96 and 3.11 and was economically feasible. At selling price of 60 Rs./kg, the BCR was more than 4 in the treatments $T_1$, $T_2$, $T_4$, $T_5$, and $T_6$ in DS-NVPH which is economically feasible. At same price in WT-NVPH, the BCR was more than 4 in the treatments $t_1$, $t_2$, $t_4$, $t_5$, and $t_6$ which are economically feasible. At the prices 80, 100, and 120 Rs./kg, the BCR ranged 3.8–6, 4.8–7, and 5.75–9.32, respectively, which were more economically feasible. In DS-NVPH, the BCR was maximum in treatment $T_5$ and minimum in the treatment $T_3$ at all the selling prices. In WT-NVPH, it was maximum in the treatment $t_2$ and minimum in $t_3$.

**TABLE 4.13** Economic Analysis of Colored Capsicum under Tensiometer-Based Irrigation Levels with Fertigation in DS-NSPH per 200 m².

| Sr. No. | Particulars | $T_1$ | $T_2$ | $T_3$ | $T_4$ | $T_5$ | $T_6$ | $T_7$ |
|---|---|---|---|---|---|---|---|---|
| 1. (i) | Fixed cost of polyhouse, Rs. | 187,000 | 187,000 | 187,000 | 187,000 | 187,000 | 187,000 | 187,000 |
| A | Life (years) | 24 | 24 | 24 | 24 | 24 | 24 | 24 |
| B | Depreciation at 10%, Rs. | 7012 | 7012 | 7012 | 7012 | 7012 | 7012 | 7012 |
| C | Interest at 13.5%, Rs. | 25,245 | 25,245 | 25,245 | 25,245 | 25,245 | 25,245 | 25,245 |
| (ii) | Fixed cost, shed net, Rs. | 28,000 | 28,000 | 28,000 | 28,000 | 28,000 | 28,000 | 28,000 |
| D | Life (years) | 4 | 4 | 4 | 4 | 4 | 4 | 4 |
| E | Depreciation at 10%, Rs. | 6300 | 6300 | 6300 | 6300 | 6300 | 6300 | 6300 |
| F | Interest at 13.5%, Rs. | 3780 | 3780 | 3780 | 3780 | 3780 | 3780 | 3780 |
| G | Total (b + c + e + f) | 42,337 | 42,337 | 42,337 | 42,337 | 42,337 | 42,337 | 42,337 |
| H | Total cost for 6 months, Rs. | 21,168 | 21,168 | 21,168 | 21,168 | 21,168 | 21,168 | 21,168 |
| 2 | Fixed cost of drip system, Rs. | 14,080 | 14,080 | 14,080 | 14,080 | 14,080 | 14,080 | 14,080 |
| A | Life (years) | 7 | 7 | 7 | 7 | 7 | 7 | 7 |
| B | Depreciation at 10%, Rs. | 1810 | 1810 | 1810 | 1810 | 1810 | 1810 | 1810 |
| C | Interest at 10%, Rs. | 1408 | 1408 | 1408 | 1408 | 1408 | 1408 | 1408 |
| D | Total (b + c), Rs. | 3218 | 3218 | 3218 | 3218 | 3218 | 3218 | 3218 |
| E | Total cost for 6 months, Rs. | 1609 | 1609 | 1609 | 1609 | 1609 | 1609 | 1609 |
| 3 | Cost of cultivation (Rs./200 m²) | 12,768 | 12,768 | 12,768 | 12,768 | 12,768 | 12,768 | 12,768 |
| 4 | Total cost for 6 months (h + e + 3), Rs. | 35,545 | 35,545 | 35,545 | 35,545 | 35,545 | 35,545 | 35,545 |
| 5 | Yield kg/200 m² | 2400 | 2468 | 2184 | 2456 | 2628 | 2508 | 1704 |

**TABLE 4.13** *(Continued)*

| Sr. No. | Particulars | $T_1$ | $T_2$ | $T_3$ | $T_4$ | $T_5$ | $T_6$ | $T_7$ |
|---|---|---|---|---|---|---|---|---|
| 6 | Selling price, Rs./kg | 20 | 20 | 20 | 20 | 20 | 20 | 20 |
| 7 | Total net seasonal income for 6 months, Rs. | 48,000 | 49,360 | 43,680 | 49,120 | 52,560 | 50,160 | 34,080 |
| 8 | B:C ratio (No.7/No.4) | 1.35 | 1.39 | 1.23 | 1.38 | 1.48 | 1.41 | 0.96 |
| 9 | Selling price, Rs./kg | 40 | 40 | 40 | 40 | 40 | 40 | 40 |
| 10 | Total net seasonal income for 6 months, Rs. | 96,000 | 98,720 | 87,360 | 98,240 | 105,120 | 100,320 | 68,160 |
| 11 | B:C ratio (No.10/No.4) | 2.70 | 2.78 | 2.46 | 2.76 | 2.96 | 2.82 | 1.92 |
| 12 | Selling price, Rs./kg | 60 | 60 | 60 | 60 | 60 | 60 | 60 |
| 13 | Total net seasonal income for 6 months, Rs. | 144,000 | 148,080 | 131,040 | 147,360 | 157,680 | 150,480 | 102,240 |
| 14 | B:C ratio (No.13/No.4) | 4.05 | 4.17 | 3.69 | 4.15 | 4.44 | 4.23 | 2.88 |
| 15 | Selling price, Rs./kg | 80 | 80 | 80 | 80 | 80 | 80 | 80 |
| 16 | Total net seasonal income for 6 months, Rs. | 192,000 | 197,440 | 174,720 | 196,480 | 210,240 | 200,640 | 136,320 |
| 17 | B:C ratio (No.16/No.4) | 5.40 | 5.55 | 4.92 | 5.53 | 5.91 | 5.64 | 3.84 |
| 18 | Selling price, Rs./kg | 100 | 100 | 100 | 100 | 100 | 100 | 100 |
| 19 | Total net seasonal income for 6 months, Rs. | 240,000 | 246,800 | 218,400 | 245,600 | 262,800 | 250,800 | 170,400 |
| 20 | B:C ratio (No.19/No.4) | 6.75 | 6.94 | 6.14 | 6.91 | 7.39 | 7.06 | 4.79 |
| 21 | Selling price, Rs./kg | 120 | 120 | 120 | 120 | 120 | 120 | 120 |
| 22 | Total net seasonal income for 6 months, Rs. | 288,000 | 296,160 | 262,080 | 294,720 | 315,360 | 300,960 | 204,480 |
| 23 | B:C ratio (No.22/No.4) | 8.10 | 8.33 | 7.37 | 8.29 | 8.87 | 8.47 | 5.75 |

$T_1$ = treatment $I_1F_1$ in DS-NVPH; $T_2$ = treatment $I_1F_2$ in DS-NVPH; $T_3$ = treatment $I_2F_1$ in DS-NVPH; $T_4$ = treatment $I_2F_2$ in DS-NVPH; $T_5$ = treatment $I_3F_1$ in DS-NVPH; $T_6$ = treatment $I_3F_2$ in DS-NVPH; $T_7$ = control treatment in DS-NVPH.

**TABLE 4.14** Economic Analysis of Colored Capsicum under Tensiometer-Based Different Levels of Irrigation and Fertigation Levels in WT-NVPH per 100 m².

| Sr. No. | Particulars | $T_1$ | $T_2$ | $T_3$ | $T_4$ | $T_5$ | $T_6$ | $T_7$ |
|---|---|---|---|---|---|---|---|---|
| 1. (i) | Fixed cost of polyhouse, Rs. | 93,500 | 93,500 | 93,500 | 93,500 | 93,500 | 93,500 | 93,500 |
| A | Life (years) | 24 | 24 | 24 | 24 | 24 | 24 | 24 |
| B | Depreciation at 10%, Rs. | 3506.25 | 3506.25 | 3506.25 | 3506.25 | 3506.25 | 3506.25 | 3506.25 |
| C | Interest at 13.5%, Rs. | 12622.5 | 12622.5 | 12622.5 | 12622.5 | 12622.5 | 12622.5 | 12622.5 |
| (ii) | Fixed cost, shed net, Rs. | 28,000 | 28,000 | 28,000 | 28,000 | 28,000 | 28,000 | 28,000 |
| D | Life (years) | 4 | 4 | 4 | 4 | 4 | 4 | 4 |
| E | Depreciation at 10%, Rs. | 6300 | 6300 | 6300 | 6300 | 6300 | 6300 | 6300 |
| F | Interest at 13.5%, Rs. | 3780 | 3780 | 3780 | 3780 | 3780 | 3780 | 3780 |
| G | Total (b + c + e + f) | 21,168 | 21,168 | 21,168 | 21,168 | 21,168 | 21,168 | 21,168 |
| H | Total cost for 6 months, Rs. | 10,584 | 10,584 | 10,584 | 10,584 | 10,584 | 10,584 | 10,584 |
| 2. | Fixed cost of drip system, Rs. | 7040 | 7040 | 7040 | 7040 | 7040 | 7040 | 7040 |
| A | Life (years) | 7 | 7 | 7 | 7 | 7 | 7 | 7 |
| B | Depreciation at 10%, Rs. | 905 | 905 | 905 | 905 | 905 | 905 | 905 |
| C | Interest at 10%, Rs. | 704 | 704 | 704 | 704 | 704 | 704 | 704 |
| D | Total (b + c), Rs. | 1609 | 1609 | 1609 | 1609 | 1609 | 1609 | 1609 |
| E | Total cost for 6 months, Rs. | 804 | 804 | 804 | 804 | 804 | 804 | 804 |
| 3. | Cost of cultivation (Rs./200 m²) | 6384 | 6384 | 6384 | 6384 | 6384 | 6384 | 6384 |
| 4. | Total cost for 6 months (h + e + 3), Rs. | 17,772 | 17,772 | 17,772 | 17,772 | 17,772 | 17,772 | 17,772 |
| 5. | Yield kg/200 m² | 1235 | 1381 | 1167 | 1349 | 1304 | 1194 | 956 |
| 6. | Selling price, Rs./kg | 20 | 20 | 20 | 20 | 20 | 20 | 20 |

**TABLE 4.14**   (Continued)

| Sr. No. | Particulars | $T_1$ | $T_2$ | $T_3$ | $T_4$ | $T_5$ | $T_6$ | $T_7$ |
|---|---|---|---|---|---|---|---|---|
| 7. | Total net seasonal income for 6 months, Rs. | 24,700 | 27,620 | 23,340 | 26,980 | 26,080 | 23,880 | 19,120 |
| 8. | B:C ratio (No.7/No.4) | 1.39 | 1.55 | 1.31 | 1.52 | 1.47 | 1.34 | 1.08 |
| 9. | Selling price, Rs./kg | 40 | 40 | 40 | 40 | 40 | 40 | 40 |
| 10. | Total net seasonal income for 6 months, Rs. | 49,400 | 55,240 | 46,680 | 53,960 | 52,160 | 47,760 | 38,240 |
| 11. | B:C ratio (No.10/No.4) | 2.78 | 3.11 | 2.63 | 3.04 | 2.93 | 2.69 | 2.15 |
| 12. | Selling price, Rs./kg | 60 | 60 | 60 | 60 | 60 | 60 | 60 |
| 13. | Total net seasonal income for 6 months, Rs. | 74,100 | 82,860 | 70,020 | 80,940 | 78,240 | 71,640 | 57,360 |
| 14. | B:C ratio (No.13/No.4) | 4.17 | 4.66 | 3.94 | 4.55 | 4.40 | 4.03 | 3.23 |
| 15. | Selling price, Rs./kg | 80 | 80 | 80 | 80 | 80 | 80 | 80 |
| 16. | Total net seasonal income for 6 months, Rs. | 98,800 | 110,480 | 93,360 | 107,920 | 104,320 | 95,520 | 76,480 |
| 17. | B:C ratio (No.16/No.4) | 5.56 | 6.22 | 5.25 | 6.07 | 5.87 | 5.37 | 4.30 |
| 18. | Selling price, Rs./kg | 100 | 100 | 100 | 100 | 100 | 100 | 100 |
| 19. | Total net seasonal income for 6 months, Rs. | 123,500 | 138,100 | 116,700 | 134,900 | 130,400 | 119,400 | 95,600 |
| 20. | B:C ratio (No.19/No.4) | 6.95 | 7.77 | 6.57 | 7.59 | 7.34 | 6.72 | 5.38 |
| 21. | Selling price, Rs./kg | 120 | 120 | 120 | 120 | 120 | 120 | 120 |
| 22. | Total net seasonal income for 6 months, Rs. | 148,200 | 165,720 | 140,040 | 161,880 | 156,480 | 143,280 | 114,720 |
| 23. | B:C ratio (No.22/No.4) | 8.34 | 9.32 | 7.88 | 9.11 | 8.80 | 8.06 | 6.46 |

$T_1$ = treatment $I_1F_1$ in WT-NVPH; $T_2$ = treatment $I_1F_2$ in WT-NVPH; $T_3$ = treatment $I_2F_1$ in WT-NVPH; $T_4$ = treatment $I_2F_2$ in WT-NVPH; $T_5$ = treatment $I_3F_1$ in WT-NVPH; $T_6$ = treatment $I_3F_2$ in WT-NVPH; $T_7$ = control treatment in WT-NVPH.

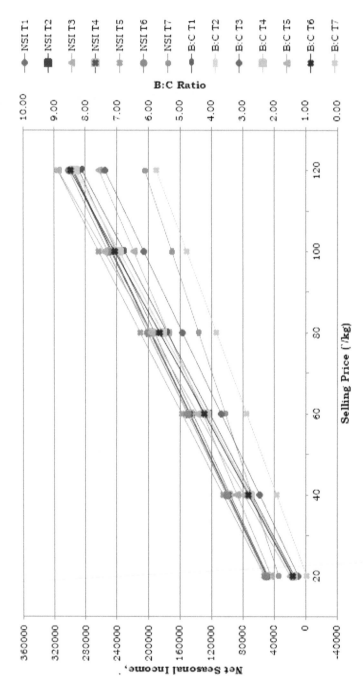

**FIGURE 4.25** Effects of market price (Rs./kg) on B:C ratio and net seasonal income on production of capsicum in DS-NVPH.

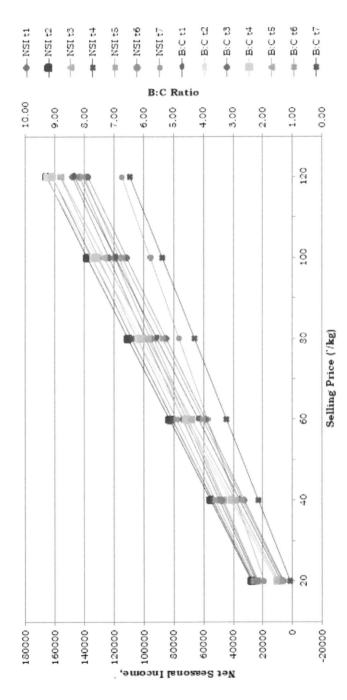

**FIGURE 4.26** Effects of market price (Rs./kg) on B:C ratio and net seasonal income on production of capsicum in WT-NVPH.

Figures 4.25 and 4.26 revealed that the production of colored capsicum under tensiometer-based drip irrigation with fertigation is economically profitable under naturally ventilated polyhouse even at 20 Rs./kg. However, the prevailing market price for the colored capsicum has been generally more than 80 Rs./kg.

### 4.4.7  SENSITIVITY ANALYSIS

The market price of colored capsicum is variable with seasonal availability (i.e., in-season and off-season). Therefore, further economic analysis was worked out to study the variable market price of colored capsicum on B:C ratio and NSI. In all cases, it was found that the B:C ratio and NSI were increased with the increase in the market price of product. The trends of different treatments in DS-NVPH and WT-NVPH (Figs. 4.25 and 4.26) indicate straight-line relationships:

$$y = ax + b, \tag{4.10}$$

where $y$ = B:C ratio or NSI (Rs.), $x$ = market price (Rs./kg), and a and b are intercept and slope constants. The values of constants $a$ and $b$ are given in Table 4.15 for DS-NVPH and WT-NVPH. The constants indicate the change in the value of B:C ratio and NSI per unit change in the market price.

**TABLE 4.15**  Regression Coefficients (a and b) for the Effects of Market Price on B:C Ratio and Net Seasonal Income in Different Treatments for DS-NVPH and WT-NVPH.

| Treatments | Regression constants | DS-NVPH | | WT-NVPH | |
|---|---|---|---|---|---|
| | | NSI | B:C ratio | NSI | B:C ratio |
| $I_1F_1$ | a | 2400 | 0.068 | 1235 | 0.069 |
| | b | 1.11E−11 | 4.9E−17 | 6.94E−12 | 3.24E−16 |
| $I_1F_2$ | a | 2468 | 0.069 | 1381 | 0.078 |
| | b | 1.69E−11 | 0.004 | 3.48E−12 | 0.000 |
| $I_2F_1$ | a | 2184 | 0.061 | 1167 | 0.066 |
| | b | 1.17E−11 | 0.005 | 1.09E−12 | −0.001 |
| $I_2F_2$ | a | 2456 | 0.069 | 1349 | 0.076 |
| | b | 7.86E−12 | −0.001 | 1.11E−11 | 0.001 |

**TABLE 4.15**   *(Continued)*

| Treatments | Regression constants | DS-NVPH | | WT-NVPH | |
|---|---|---|---|---|---|
| | | NSI | B:C ratio | NSI | B:C ratio |
| $I_3F_1$ | a | 2628 | 0.074 | 1304 | 0.073 |
| | b | 1.95E−11 | 0.004 | 1.13E−11 | −5E−16 |
| $I_3F_2$ | a | 2508 | 0.071 | 1194 | 0.067 |
| | b | 1.07E−11 | −0.005 | 1.01E−12 | −0.001 |
| C | a | 1704 | 0.048 | 956 | 0.054 |
| | b | 4.04E−12 | 0.005 | 7.77E−12 | 0.001 |

## 4.5   CONCLUSIONS

Greenhouse technology has tremendous potential in keeping up the pace of agriculture. Protected cultivation of vegetables and flowers not only increases the sustainability of agricultural production but also improves the standard of living and contribution of agriculture in GDP. Greenhouse technology is the technique of providing favorable environment condition to the plants. It is rather used to protect the plants from the adverse climatic conditions such as wind, cold, precipitation, excessive radiation, extreme temperature, insects, and diseases. It is also of vital importance to create an ideal microclimate around the plants. This is possible by erecting a greenhouse/polyhouse, where the environmental conditions are so modified that one can grow any plant in any place at any time by providing suitable environmental conditions with minimum labor.

In the present scenario of perpetual demand of flowers, vegetables, orchards, and shrinking land holding drastically, greenhouse cultivation is the best alternative for using land water and other resources more efficiently. Greenhouse plays an important role for cultivating off-season vegetables, which elbow out the constraints of limited fertile land, environmental pollution, toxic crop residues, and seasonality of vegetables in relation to various agricultural regions and natural climatic upsets.

High-quality fruits and vegetables grown under micro irrigation in a protected environment (greenhouse/polyhouse) can be found these days in the most exclusive supermarkets and flower stands around the globe. It is employment-generating technology and provides more foreign exchange due to export.

It is, therefore, clear that vegetable production can become sustainable only if we judiciously use our costly and limited water and land resources with the help of modern science and technology.

For optimum to maximum production of vegetables, water is the most important factor with the greenhouse environment. As we utilize minimum required space for greenhouse cultivation same as if we irrigate accurate amount of water required by the plant that is not possible through micro/drip irrigation system alone. For this purpose, scheduling of irrigation is necessary. With the different stages of growth crop water requirement also be different so, for every stage of crop calculated amount of water should be given on the basis of plant-soil moisture tension with help of tensiometer instrument, which is the easiest alternative option than other methods.

The main purpose of this chapter is to investigate the tensiometer-based drip irrigation scheduling of capsicum (*Capsicum annum* L.) under polyhouse in order to increase WUE, and to maximize the net returns. Following conclusions were drawn:

1.  From this study, it can be concluded that the optimal temperature for the growth of capsicum crop was 16°C–25°C because fruit-set does not occur below 16°C, and above 25°C there is flower abscission.

2.  During the 6-month growing period in DS-NVPH, the average depth of water applied at the irrigation levels $I_1$ (20–30 kPa) was 249.06 mm, $I_2$ (30–50 kPa) was 174.06 mm, $I_3$ (50–70 kPa) level was 130.62 mm, and in control C was 322.98 mm. In WT-NVPH, the average depth of water applied at the irrigation level $I_1$ (20–30 kPa) was 303.27 mm, $I_2$ (30–50 kPa) was 190.93 mm, at $I_3$ (50–70 kPa) level was 143.47 mm, and in C was 366.27 mm.

3.  The mean water productivity for capsicum at the tensiometer-based irrigation levels $I_1$, $I_2$, $I_3$ and in control, in DS-NVPH was 20.55, 15.74, 10.31, and 37.98 L/kg, respectively. And in WT-NVPH, it was 23.31, 13.57, 11.54, and 38.53 L/kg, respectively.

4.  The mean yield of capsicum in DS-NVPH at irrigation levels $I_1$, $I_2$, and $I_3$ was 12.17, 11.60, 12.84 kg/m$^2$, and in control 8.52 kg/m$^2$. And in WT-NVPH it was 13.80, 12.58, 12.49 kg/m$^2$, and 9.56 kg/m$^2$, respectively.

5.  In DS-NVPH and WT-NVPH, values of biometric parameters, the yield and yield attributes were higher in $I_1$ level of irrigation and followed by $I_2$ and $I_3$.

6.  Production of colored capsicum under tensiometer-based different irrigation levels and different fertigation levels in both polyhouses was economically feasible even at the selling prices from 20 Rs./ kg and highly profitable for the grower at the prevailing market price (80–250 Rs./kg).

## 4.6  SUMMARY

The present study was undertaken to investigate the "tensiometer-based drip irrigation scheduling of capsicum (*Capsicum annum* L.) under polyhouse". The site was located at experimental field of Department of Irrigation and Drainage Engineering, College of Technology, G. B. Pant University of Agriculture and Technology, Pantnagar. The study included research on various meteorological parameters inside the polyhouses, the response of tensiometer-based irrigation levels in combination with fertigation levels on the biometric growth and yield parameters of capsicum, and economic feasibility of colored capsicum production under the polyhouses.

The mean maximum temperature was higher in WT-NVPH (34.07°C) than the DS-NVPH (33.06°C). The minimum temperature was also higher in the WT-NVPH (19.83°C) than that in DS-NVPH (18.97°C).

The mean daily variation of temperature at 9:00 am was higher in WT-NVPH (15.73°C) than DS-NVPH (15.30°C). At 12:00 noon and 16:00 p.m., it was also higher in WT-NVPH than DS-NVPH. The mean daily variation in relative humidity at 9:00 a.m., 12:00 noon, and 16:00 p.m. was higher in WT-NVPH than DS-NVPH. Similarly, mean daily solar intensity at these hours of the day was also higher in WT-NVPH than DS-NVPH. The crop evapotranspiration ($ET_c$) was higher in WT-NVPH than DS-NVPH. The average mean daily $ET_c$ in WT-NVPH and DS-NVPH was 2.22 and 2.10 mm/day, respectively. During the 6-month growing period in DS-NVPH, the average depth of water application at the irrigation level $I_1$ (20–30 kPa) was 249.06 mm, $I_2$ (30–50 kPa) was 174.06 mm, and at $I_3$ (50–70 kPa) level was 130.62 mm, whereas depth of irrigation was 322.98 mm in conventional practice (control). In WT-NVPH, the average depth of water at the irrigation level $I_1$ was 303.27 mm followed by $I_2$ (190.93 mm), and $I_3$ (143.47 mm) whereas in control it was 366.27 mm.

Production of the mean value of capsicum in DS-NVPH at irrigation levels $I_1$, $I_2$, $I_3$, and control was 12.17, 11.60, 12.84, and 8.52 kg/m² while in WT-NVPH it was 13.80, 12.58, 12.49, and 9.56 kg/m$^2$, respectively. The average water productivity for capsicum at irrigation levels $I_1$, $I_2$, $I_3$, and control in DS-NVPH was 20.55, 15.74, 10.31, and 37.98 L/kg, respectively. Further, in WT-NVPH water productivity was 23.31, 13.57, 11.54, and 38.53 L/kg, respectively. Production of colored capsicum under tensiometer-based irrigation levels and in control treatments in both the polyhouses were economically feasible at selling price higher than 20 Rs./kg.

## KEYWORDS

- benefit: cost ratio
- crop evapotranspiration
- fertigation
- tensiometer
- yield

## REFERENCES

1. Abbott, J. D.; Peet, M. M.; Willits, D. H.; Sanders, D. C.; Gough, R. E. Effects of Irrigation Frequency and Scheduling on Fruit Production and Radial Fruit Cracking in Greenhouse Tomatoes in Soil Beds and in a Soil-less Media in Bags. *Sci. Hortic.* **1985,** *28*(3), 209–217.
2. Agarwal, K. N.; Satapathy, K. K. *Potential of Using Low Cost Polyhouse in NEH Region.* Proceedings of All India Seminar on Potential and Prospects for Protective Cultivation organized by the Institute of Engineers, Ahmednagar, December 12–13, 2003; pp 47–53.
3. Allen, R. G., Pereira, L. S., Raes, D.; Smith, M. (1998). *Crop Evapotranspiration: Guidelines for Computing Crop Water Requirements.* Irrigation and Drainage Paper 56. Food and Agriculture Organization of the United Nations (FAO): Rome, Italy.
4. Anitha, S.; Jyothi, M. L.; Narayanan-Kutti, M. C.; Nari, L. Evaluation of Various Organic Manures as Components in The Integrated Nutrient Management of Oriental Pickling Melon *(Cucumis melo* var. *conomon). Prog. Hort.* **2003,** *35*, 155–157.
5. Aviscor, R.; Mahrer, Y. Verification Study of Numerical Greenhouse Microclimate Model. *Trans. ASAE.* **1982,** *25*(6), 1711–1720.
6. Blanco, F. F.; Folegatti, V. (2002). Nitrogen and potassium effects on tomato salinity tolerance in greenhouse. ASAE Paper No. 022188 St. Joseph, Mich.: ASAE, USA.

7. Bonachela, S.; Gonzalez, A. M.; Fernandez, M. D. Irrigation Scheduling of Plastic Greenhouse Vegetable Crops Based on Historical Weather Data. *Irr. Sci.* **2006,** *25,* 53–62.

8. Brar, G. S.; Varshenya, M. C.; Sable, R. N.; Salunkhe, S. S.; Hazari, A. K. Influence of Irrigation and Light Levels on Transpiration in Capsicum under Polyhouse. *J. Agrometerol.* **2006,** *8*(2), 192–196.

9. Chartzoulakis, K., Drosos, N.; Chartzoulakis, K. S. Water requirement of greenhouse grown pepper under drip irrigation. Proceedings of the Second International Symposium on Irrigation of Horticultural crops, Chania, Crete, Greece, 9–13 September. *Acta Horti.* **1997,** *449,* 175–180.

10. Chiew, F. H. S., Kamaldasa, N. N., Malono, H. M.; Memohon, T. A. Penman-Monteith, Reference Crop Evapotranspiration and Class A Pan Data in Australia: FAO 24. *Agric. Water Manag.* **1995,** *28,* 9–21.

11. Dayan, E.; Enoch, H. Z.; Fuchs, M.; Zipori, I. Suitability of Greenhouse Building Types and Roof Cover Materials for Growth of Export Tomatoes in the Bensor Region of Israel, I: Effect on Climatic Conditions. *Biotronics, 15,* 61–70.

12. Demirtaş, C.; Ayas, S. Deficit Irrigation Effects on Pepper (*Capsicum annuum* L. Demre) Yield in Unheated Greenhouse Condition. *J. Food Agric. Environ.* **2009,** *7*(3&4), 989–993.

13. Deolankar, K. P.; Firake, N. N. Effect of Fertigation on Solid Soluble Fertilizers on Growth and Yield of Chilli. *J. Maharashtra Agric. Univ.* **1999,** *24*(3), 242–243.

14. Desai, A. Vermiculture Application in Horticulture. In Congress Proceedings on Traditional Science and Technology: Key Notepapers and Extended Abstracts, IIT, Bombay, 1992, 10, 10–11.

15. Doorenbos, J.; Pruitt, W. O. (1977). *Guidance for Predicting Crop Water Requirements.* Irrigation and Drainage Paper 24. United Nations FAO: Rome, Italy.

16. Fceilla, A.; Cascone, G. The Energy Efficiency of Structures Covered with Double Plastic Films. *Colture-Protette.* **1987,** *16*(2), 57–62.

17. Samajstrla, G. Irrigation Scheduling of Drip Irrigated Tomato Using Tensiometer and Pan Evaporation. *Proc. Fla. State Hortic. Soc.* **1990,** *103,* 88–91.

18. Gajanana, T. M.; Singh, K. P.; Subrahamnyam, K. V.; Mandhari, S. C. Economic Analysis of Gerbera Cultivation under Protected Cultivation. *Indian J. Hortic.* **2003,** *60,* 104–107.

19. Gercek, S.; Comlekcioglu, N.; Dikilitas, M. Effectiveness of Water Pillow Irrigation Method on Yield and Water Use Efficiency on Hot Pepper. *Sci. Hortic.* **2009,** *120,* 325–329.

20. Gomez, H. V.; Farias, S. O.; Argote, M. Evaluation of Water Requirements for a Greenhouse Tomato Crop Using the Priestley-Taylor Method. *Chil. J. Agr. Res.* **2007,** *69*(1), 3–11.

21. Gonzalez, A. M.; Bonachela, S.; Fernandez, M. D. Regulated Deficit Irrigation in Green Bean and Watermelon Greenhouse Crops. *Sci. Hortic.* **2009,** *122,* 527–531.

22. Goswami, W.; Mohammad, M. J.; Najim, H.; Qubursi, R. Response of Bell Pepper Grown Inside Plastic Houses to Nitrogen Fertilization. *Commun. Soil Sci. Plan.* **1999,** *30*(17/18), 2499–2509.

23. Goyal, M. R. *Research Advances in Sustainable Drip Irrigation, Volumes 1 to 10*; Apple Academic Press Inc.: Oakville, ON, 2015.

24. Goyal, M. R. *Challenges and Innovations in Micro Irrigation, Volumes 1 to 5;* Apple Academic Press Inc.: Oakville, ON, 2016.
25. Graber, E. R.; Harel, Y. M.; Kolton, M.; Cytryn, E.; Silber, A.; David, D. R.; Tsechansky, L.; Borenshtein, M.; Elad, Y. Biochar Impact on Development and Productivity of Pepper and Tomato Grown in Fertigated Soilless Media. *Springer Sci. Plant Soil.* **2010**, 1–16.
26. Graverend, G. Le. The Solar-Heated Greenhouse. *Hortic. Francaise.* **1976**, *68*, 17–18.
27. Hargreaves, G. H. Defining and Using Reference Evapotranspiration. *J. Irrig. Drain. Eng. ASCE.* **1984**, *120*(6), 1132–1139.
28. Hargreaves, G. H.; Samani, Z. A. Reference Evapotranspiration from Temperature. *Appl. Eng. Agric.* **1985**, *1*(2), 96–99.
29. Harmanto, S. V. M.; Babel, M. S.; Tantau, H. J. Water Requirement of Drip Irrigated Tomatoes Grown in Greenhouse Tropical Environment. *Agr. Water Manage.* **2005**, *71*(3), 225–242.
30. Hensley, D.; Deputy, J. (1999). *Using Tensiometers for Measuring Soil Water and Scheduling Irrigation.* Cooperative Extension Service, College of Tropical Agriculture and Human Resources (CTAHR), University of Hawaii at Manoa, Lecture 10, pp 1–4.
31. Hills, D. J.; Brenes, M. J. Micro-irrigation of Waste Water Effluent Using Drip Tape. *Appl. Eng. Agric.* **2001**, *17*(3), 303–308.
32. Hoppula, I. K.; Salo, J. T. Tensiometer-based Irrigation Scheduling with Different Fertilization Methods in Blackcurrant Cultivation. *Acta Agric. Scand. Sect. B-Soil Plant.* **2005**, *55*, 229–235.
33. Hussein, A. S. A.; Eldraw, A. K. Evapotranspiration in Sudan Gezira Irrigation Scheme. *J. Irrig. Drain. Eng. ASCE.* **1989**, *115*(6), 1018–1033.
34. Iqbal, M.; Khatri, A. K. Wind Coefficient from Long Semicircular Greenhouses. *Trans. ASAE.* **1976**, 911–914.
35. Jeevansab, C. Effect of Nutrient Sources on Growth, Yield and Quality of Capsicum Grown under Different Environments. M.Sc. (Agri) Thesis, University of Agricultural Sciences, Dharwad, 2000.
36. Khan, M. H.; Chattha, T. H.; Nadia, S. Influence of Different Irrigation Intervals on Growth and Yield of Bell Pepper. *Res. J. Agric. Biol. Sci.* **2005**, *1*(2), 125–128.
37. Khurana, D. S.; Singh, R.; Sidhu, A. S.; Ranjodh, S. Effect of Different Levels of Nitrogen in Split Doses on Growth and Yield of Chilli. *Indian J. Hortic.* **2006**, *63*(4), 467–469.
38. Kittas, C.; Boulard T.; Bartznas, T.; Katsoulas, N.; Mermier, M. Influence of an Insect Screen on Greenhouse Ventilation. *Trans. ASAE.* **2002**, *45*(4), 1083–1090.
39. Kittas, C.; Katsoulas, N.; Baille, A. Influence of an Aluminized Thermal Screen on Greenhouse Microclimate and Canopy Energy Balance. *Trans. of the ASAE*, **2003** *46*(6), 1653–1663.
40. Kittas, C.; Katsoulas, N.; Baille, A. Influence of an Aluminized Thermal Screen on Greenhouse Microclimate and Canopy Energy Balance. *Trans. of the ASAE*, **2003** *46*(6), 1653–1663.
41. Klocke, N. L.; Martin, D. L.; Todd, R. W.; Dettaan, D. L. Evaporation Measurements and Predicting from Soil under Crop Canopies. *Trans. ASAE.* **1990**, *13*(5), 1590–1596.

42. Klosowski, E. S.; Lunardi, D. M. C. Red Pepper (*Capsicum annum* L.) Crop Coefficient Determination Cultivated in Greenhouse Conditions. *Energ. Agri.* **2002**, *17*(4), 33–42.

43. Krueger, E.; Schmidt, G.; Brueckner, U. Scheduling Strawberry Irrigation Based upon Tensiometer Measurement and a Climatic Water Balance Model. *Sci. Horti.*, **1999**, *81*, 409–424.

44. Casanova, M. P.; Messing, I., Joel, A.; Canete, A. (2008). Methods to Estimate Lettuce Evapotranspiration in Greenhouse Conditions in the Central Zone of Chile. *Chil. J. Agr. Res.* **2008**, *69*(1), 60–70.

45. Mahajan, G.; Singh, K. G. Response of Greenhouse Tomato to Irrigation and Fertigation. *Agric. Water Manag.* **2006**, *84*, 202–206.

46. Manjunath, B. L.; Mishra, P. K.; Rao, J. V.; Reddy, G. S. Water Requirement of Vegetables in a Dry Land Watershed. *Indian J. Agr. Sci.*, **1994**, *64*(12), 845–846.

47. Monteith, J. L. Evaporation and Surface Temperature. *Quart. J. Roy. Meteor. Soc.* **1981**, *107*, 1–27.

48. Mpusia, P. T. O.; Su, Z.; Becht, R.; Bos, M. G. Comparison of Water Consumption between Greenhouse and Outdoor Cultivation. M.Sc. Thesis, International Institute for Geo-information Science and Earth Observation Enschede, the Netherlands, 2006.

49. Murthy, D. S.; Prabhakar, B. S.; Hebbar, S. S.; Srinivas, V.; Prabhakar M. Economic Feasibility of Vegetable Production under Polyhouse: A Case Study of Capsicum and Tomato. *J. Hortic. Sci.* **2009**, *4*(2), 148–152.

50. Nicole L. S.; George, J. H.; Hanlon, E. A. *N Fertigation Management for Drip Irrigated Bell Pepper (Capsicum annuum L.)*. Proceedings of the Florida State Horticultural Society, *109*, 136–141, 1996.

51. Olczyk, T.; Regalado, R.; Li, Y.; Jordan, R. *Usefulness of Tensiometers for Scheduling Irrigation for Tomatoes Grown on Rocky, Calcareous Soils in Southern Florida.* Proceedings of the Florida State Horticultural Society, *113*, 239–242, 2000.

52. Parvej, M. R.; Khan, M. A. H.; Awal, M. A. Phenological Development and Production Potentials of Tomato under Polyhouse Climate. *J. Agri. Sci.* **2010** *59*(1), 19–31.

53. Pereira, L. S.; Perrier, A.; Allen, R. G.; Alves, I. Evapotranspiration: Concepts and Future Trends. *J. Irrig. Drain. E. Div., ASCE.* **1999**, *125*(2), 45–51.

54. Prenger, J. J.; Fynn, R. P.; Hansen, R. C. A Comparison of Four Evapotranspiration Models in a Greenhouse Environment. *Trans. ASAE.* **2002**, *45*(6), 1779–1788.

55. Samani, Z. Estimating Solar Radiation and Evapotranspiration using Minimum Climatological Data. *J. Irrig. Drain. Engg.* **2000**, *126*(4), 265–267.

56. Samsuri, S. F. M.; Ahmad, R.; Hussein, M.; Rahim, A. R. A. *Identification and Modeling of Environmental Climates Inside Naturally Ventilated Tropical Greenhouse.* Fourth Asia International Conference on Mathematical/Analytical Modelling and Computer Simulation, University Technology Johor, Malaysia, 2010; pp 402–407.

57. Sarker, J. U.; Halder, N. K. Response of Chilli to Integrated Fertilizer Management in North-Eastern Brown Hill Soils of Bangladesh. *J. Biol. Sci.* **2003**, *3*(9), 797–801 (online).

58. Sharma, B. R. Water Management in the 21st Century: Priority Issues. *J. Water Manag.* **2002**, *10*(1/2), 11.

59. Shih, S. F. Data Requirement for Evaporation Estimation. *J. Irrig. Drain. Engg. ASCE.* **1983**, *110*(3), 763–274.

60. Shinde, P. P.; Ramteke, J. R.; More, V.G.; Chavan, S. A. Evaluation of Micro Irrigation Systems and Mulch for Summer Chilli Production. *J. Maharashtra Agric. Univ.* **2002,** *27*(1), 51–54.

61. Shinde, U. R.; Firake, N. N. *Economics of Summer Chilli Production with Mulching and Micro-Irrigation*; MPKV: Rahuri, India, 1997.

62. Shoemaker, J. S.; Taskey, B. J. E. *Practical Horticulture.* John Wiley and Sons Inc.: New York, 1995.

63. Silveston, P. L. Energy Conservation through Control of Greenhouse Humidity. I. Condensation of Heat Losses. *Can. Agr. Engg.* **1980,** *22*, 125–132.

64. Singh, R.; Asrey, R.; Nangare, D. D. *Studies on the Performance of Tomato and Capsicum under Medium Cost Greenhouse.* Proceedings of All India Seminar on Potential and Prospects for Protective Cultivation, organized by the Institute of Engineers, Ahmednagar, December 12–13, 2003; pp 158–161.

65. Smith, M.; Allen, R. G.; Monteith, J. L.; Perrier, A.; Pereira L.; Segren A. *Report of the Expert Consultation on Procedure for Revision of FAO Guidelines for Prediction of Crop Water Requirements*; UN-FAO, 54 pages, 1992.

66. Sondge, V. D.; Rodge R. P. Response and Economic Analysis of Water and Nitrogen to Chilli. *J. Maharashtra Agric. Univ.* **1991,** *16*(1), 10–12.

67. Sriharsha, B. Influence of Environmental Factors, Nutrient Sources on Growth, Yield and Quality of Tomato. M.Sc. (Agri) Thesis, University of Agricultural Sciences, Dharwad, 2001.

68. Sumarna, A. and Kusandriani. The Effect of Water Supply Amount on the Growth and Yield of Sweet Pepper (*Capsicum annum* L.) Var. grossum c.v. Orion Yolo wonder A. *Bulle. Peneli. Horti.* (Indonesia) **1992,** *24*(1) 51–58.

69. Takakura, T.; Kubotab, C.; Sasea, S.; Hayashic, M.; Ishiia, M.; Takayamad, K.; Nishinad, H.; Kuratae, K.; Giacomelli, G. A. Measurement of Evapotranspiration Rate in a Single-Span Greenhouse using the Energy-Balance Equation. *Bio. Sys. Engg.* **2008,** I (*2*), 298–304.

70. Thorat, P. V.; Balwan, A. D.; Ahire, D. D. *Effect of Protective Cover (shed net) with Different Levels of Fertilizers and Irrigation on Growth and Yield of Capsicum*; Progress Report, CAET: Dapoli, India, 2007.

71. Tumbare, A. D.; Bhoite, S. U.; Nikam, D. R. Effect of Planting Techniques and Apportioning Fertilizer Through Fertigation on the Productivity and Nutrient Uptake of Summer Chilli. *J. Maharashtra Agric. Univ.* **2007,** *32*(3), 316–318.

72. Wang, Y.; Zhang, Y. Effect of Greenhouse Subsurface Irrigation on Soil Phosphatase Activity. *Comm. Soil Sci. Plant Anal.* **2008,** *39*, 680–692.

73. Watanbe, F.; Takahashi, H., Ismael, T. M.; Takahashi, S. Estimation of Irrigation Water Requirement in the Republic of Djibouti. *J. Agri. Dept. Stud.* (Japan) **2001,** *12*(1), 40–46.

74. Willits, D. H.; Li, S. A Comparison of Naturally Ventilated vs. Fan Ventilated Greenhouse in the Southeastern United States. ASAE Paper 054155 St. Joseph, Mich.: ASAE, USA, 2005.

75. Yoon, J. Y.; Green, S. K.; Tschanz, A. T.; Tsou, S. C. S.; Changa, L. C. *Pepper Improvement for Tropics: Problems and the AVRDC Approach in Tomato and Pepper Production in the Tropics.* International Symposium on Integrated Management Practices; pp 86–98.

76. Yuan, B. Z., Yaohu Kang, Y. and Zhu, O. Drip Irrigation Scheduling for Tomatoes in Unheated Greenhouse. In *Ecosystem Service and Sustainable Watershed Management in North China*, International Conference, Beijing, P. R. China, August 23–25, 2000; pp 506–518.

77. Zhang, P.; Wang, Y. K.; Zhan, J. W.; Wang, X.; Wu, P. Scheduling Irrigation for Jujube (*Ziziphus Jujuba Mill.*). *African J. Biotechnol.* **2010**, *9*(35), 5694–5703.

# CHAPTER 5

# OPEN AND COVERED CULTIVATION: IRRIGATION AND FERTIGATION SCHEDULING

P. K. SINGH

*Department of Irrigation and Drainage Engineering, College of Technology, G. B. Pant University of Agriculture and Technology, Pantnagar 363145, Uttarakhand, India*

*E-mail: singhpk67@gmail.com*

## CONTENTS

## ABSTRACT

Precise irrigation and fertigation allow an accurate and uniform application of water and nutrients to the wetted area, where the active roots are concentrated. Planning the irrigation system and nutrient supply to the crops according to their physiological stage of development, and consideration of the soil and climate characteristics, result in high yields and high-quality crops with minimum pollution. In this chapter, efforts were made to discuss various issues of irrigation and fertigation scheduling for the precision water and nutrient management for achieving high nutrient use efficiency under protected and open cultivation.

## 5.1   INTRODUCTION

Significant progress has been made in irrigation and fertilizer application technologies and in the implementation of water and nutrient management practices such as: scientific irrigation and fertilizer scheduling under conventional and modern irrigation techniques. However, scientists report that irrigation inefficiency remains the rule rather than the exception. Gains in water and nutrient use efficiency can be achieved when water application and scheduling is precisely matched to the site-specific (spatially distributed) crop demand that is a central principle underlying precision irrigation and fertigation. This site-specific crop water demand is present in agricultural fields mainly because of variability in soil properties and topography but may also result from variable rainfall or crop variation associated with multiple crops planted in the same field or plants growing at different phonological stages induced by natural or manmade causes. There are many examples of precision irrigation over the last two decades in the western world particularly in the USA and also in Israel. In India and other developing countries, the term precision irrigation means efficient methods of water application through sprinkler and drip irrigation. However, few systems of micro irrigation and sprinkler irrigation have been installed in automated mode based on time, volume, and real-time soil moisture feedback system for efficient irrigation scheduling.

Scheduling of nutrients at right time, in right amount, in right manner, and at right place is the crux of precision nutrient management. Micro irrigation, a technique that provides crops with water through a network of pipe lines at a high frequency but with a low volume of water (drips) applied directly to the root zone in a quantity that approaches consumptive use of the plants, can be combined with fertilizer application, to offer fertigation. Fertigation enables the farmer to meet the specific water and nutrient needs of the crop with great precision, thus minimizing losses of both precious water and nutrients. The direct delivery of fertilizers through drip irrigation demands the use of soluble fertilizers and pumping and injection systems for introducing the fertilizers directly into the irrigation system [8]. Fertigation allows an accurate and uniform application of nutrients to the wetted area, where the active roots are concentrated. The nutrients are applied as per the crop need at different growth stages in split manner. The problem of mobility of non-mobile nutrients is also addressed using fertigation.

Planning the irrigation system and nutrient supply to the crops according to their physiological stage of development, and consideration of the soil and climate characteristics, results in high yield and high-quality crop with minimum pollution. In India, more than 4.0 Mha of land have been brought under pressurized irrigation (sprinkler and micro irrigation). Most of the crops irrigated under micro irrigation are horticultural crops. However, field crops such as sugarcane, groundnut, cotton, etc. are also being brought under micro irrigation. Efforts are going on to develop economical design of micro-fertigation system for the efficient water and nutrient management in crops like paddy and wheat. The fertilizer applications to these micro irrigated crops are partially through fertigation in open field conditions. However, most of the crops under polyhouse conditions are under fertigation.

In this chapter, efforts have been made to discuss various issues of fertigation for the precision nutrient management for achieving high nutrient use efficiency.

## 5.2  IRRIGATION TECHNIQUES FOR PROTECTED (GREEN HOUSE/SHADE NET HOUSE AND WALKING/LOW TUNNEL) AND OPEN FIELD CULTIVATION

It is very difficult to obtain full benefits from protected cultivation without micro irrigation and fertigation. An efficient irrigation system, preferably micro irrigation, combined with fertigation system must be an essential component of protected cultivation. Therefore, an irrigation system is essential for growing plants in a greenhouse. Plants rely on water to live and grow and because a greenhouse will not allow natural rainfall in, artificial means for irrigation become necessary. A variety of irrigation methods exist, and each method has its benefits and drawbacks. Choosing the best irrigation method depends largely on the size of the protected structure and the types of plants growing inside. Often the most effective irrigation comes from a combination of methods.

## 5.2.1  TYPES OF IRRIGATION SYSTEMS FOR NURSERIES

A plant nursery requires an irrigation system to supply water to plants effectively and simultaneously. Creating a nursery to house the plants can involve a significant effort and commitment. Necessary decisions include selecting plant species, soil types and building materials, and the amount of time and money available to invest in such a project. Nurserymen should think about which watering or irrigation system best suits their nursery type and size.

### 5.2.1.1  OVERHEAD SPRINKLER IRRIGATION SYSTEM

Nurserymen using overhead sprinklers typically have two options. The first option, rotary sprinkler heads, contain a rotating nozzle that sends a torrent of water over plants. The second option, stationary sprinkler heads, sends a rapid flow of water against a plate. The impact disrupts the steady stream of water, turning it into a continuous spray that waters plants.

Although overhead sprinkler systems are the most common option in nurseries, they are not very efficient. They require a high-pressure pump that consumes large quantity of energy. Overhead sprinklers also waste about 80% of the water emitted. In nurseries containing plants with large or broad leaves, these plants encourage water waste when leaves redirect water away from plant containers rather than into the soil. Some gardeners compensate for water loss by installing slanted plant beds that channel water into ponds where it can accumulate and be recycled back into the nursery, although this may also recycle bacteria, sodium, fertilizer, or pathogens as well.

## 5.2.1.2   MICRO IRRIGATION SYSTEM

Unlike overhead sprinklers, micro irrigation systems (drip or micro sprinkler) are highly efficient and can function using low pressure. However, soil, algae, and chemical fertilizers can clog emitters for which various types of filters are provided. Three types of micro irrigation systems are used in nurseries. One type of micro irrigation, known as the capillary mat system, uses tubes that carry water into a mat. The mat becomes saturated with water, providing containers sitting on top of the mat with a supply of water to soak up through plant root systems. Although capillary mat irrigation uses 60% less water than conventional overhead sprinkler systems, they can cause salt accumulation in the soil over long periods of time.

The second type of micro irrigation system is known as a micro sprayer, micro sprinkler, or spray stake system. Considered one of the most efficient nursery irrigation systems, micro sprayers use a tube to carry water directly into the soil from a water source. Not only does this eliminate water waste that is deflected off broad plant leaves, micro sprayers carry water directly to the plant's root system. Although micro sprayers cost more than overhead sprinklers when installed in small plants, yet they operate efficiently in larger plants with more foliage and heavier canopies.

The third type of micro irrigation is known as the spaghetti tube system. This nursery irrigation method uses narrow tubes to bring water into the plant container. A miniature weight at one end of the tube ensures that it stays in the container. Water travels from one pore to another, through a capillary system. Consequently, gardeners must use a high-quality, uniform soil for maximum efficiency. When using the spaghetti tube system, gardeners should keep soil moist at all times, as dry soil will lead to poor water distribution.

## 5.2.1.3   CAPILLARY SAND BEDS

Unlike sprinkler and micro irrigation systems, capillary sand-beds do not involve any electricity. Containing wood panels, a plastic liner, sand, a small water reservoir, drainage pipe, and valve, capillary sand-beds are built to slant slightly, allowing water released into the raised end to slowly travel to the lower end. Providing an even and continuous water supply, capillary sand-beds involve less maintenance. Plants grow evenly, relying less on fertilizer and pesticide. However, capillary sand-beds do attract weeds. Gardeners can purchase products to reduce the occurrence of weeds or they can remove them manually. Capillary sand-beds also have high installation costs.

### 5.2.1.4  MIST IRRIGATION SYSTEMS RECOMMENDED IN POLYHOUSE

Overhead irrigation can produce constant humidity in polytunnels.A mist sprinkler system produces a very fine spray or mist over plants. Greenhouses, polyhouses, and polytunnels can use either overhead or bench misting. The best type for specific need depends on: plant type, the size of polyhouse essentially plastic over a framework, and on the growing conditions in the area.

#### 5.2.1.4.1  Overhead Systems

In overhead misting systems, lines or sprinklers are installed under the roof framework of the polyhouse, and this "rain" water down onto the plants. This type of irrigation system is easy to automate and produces high humidity. This high humidity allows to protect crops against frost damage. For the best coverage, space overhead sprinklers are around 50–60% of the wetting diameter of the sprinkler.

#### 5.2.1.4.2  Bench Misting

Bench misting uses a central line of sprinklers or hoops placed at or just above the level of the plants. Bench misting requires plants to be placed on raised benches, and these must be made of materials that are impervious to water, such as a metal. One can also use a self-contained misting bench, in which the bench is partially covered with a "roof." This allows one to have just a single misting bench in the polyhouse, and a different watering system in the rest of the polyhouse.

Both types of misting system are well suited to plants that need to be kept moist, such as seedlings, and to reducing temperatures in a polyhouse. Misting is commonly used for propagation and for growing tropical plants that require constant humidity. Some misting systems can also be used to spray fertilizers evenly and finely. Fertilizer applied this way is more easily absorbed into the plants than fertilizer applied on the soil. By allowing to vary the humidity within the polyhouse, mist systems also allow to vary the temperature and can control the growing conditions.

It can be easy to over-water with either type of mist irrigation system. To prevent this, one can use a timer to turn the water on and off. Misting nozzles have very fine holes that can clog up if hard water is used. Misting also works best if polyhouse is completely enclosed, as a breeze can disrupt the fine spray and cause areas to remain dry. Misting may not be suitable for all types of plants, so if polyhouse has many different types of plants, with different water tolerances, one may need to water each type of plant individually, or use individual benches, rather than use an overhead mist irrigation system.

## 5.2.2   DRIP IRRIGATION SYSTEM

Drip irrigation is one of the most efficient methods of watering, typically operating at a 90% efficiency rate. Runoff and evaporation are minimum when compared to other irrigation systems. In drip irrigation, tubes that have emitters run alongside the plants receiving irrigation. The water leaves the tubing through the emitters by slowly dripping into the soil near the root zone. This method of irrigation minimizes leaf, fruit, and stem contact with water resulting in reduced plant disease. It reduces weed growth by keeping the area between plants dry. Irrigation through the drip method can be set to run automatically or controlled manually.

Drip watering is an excellent way to supply water to plants in greenhouses and tunnels as it keeps the humidity low leading to less pest and disease problems. Water is directed to exactly where it is needed either with an individual dripper, especially for pot-grown plants, or inserted into a pipe for beds. It is ideal for raised beds.

## 5.3   IRRIGATION SCHEDULING

### 5.3.1   GREENHOUSE DRIP IRRIGATION SCHEDULING

Availability and decreased quantity of water for agriculture highlights the objective of optimizing productivity, with adequate and efficient irrigation, that replenish the root zones soil water deficit and maximize the applied water that is stored in the rooted soil profile and used afterwards by the crop, in order to reach best yields [4]. As crop responds more to soil moisture content and irrigation regime than to method of irrigation, information developed for other irrigation methods is applicable to drip systems.

Four basic questions must be answered to pursue precise irrigation scheduling:

a.  When to irrigate?: Frequency.
b.  How much water to apply?: The amount of water to be applied must replenish the evapotranspired water, once corrected by the application efficiency (as far as the soil-water content variations are unimportant, due to the high frequency of drip irrigation). When saline water is used, the applied water must cover the leaching requirements [1, 7, 21, 22]. Other components of the water balance are normally unimportant in drip-irrigated greenhouses (unless the rainfall penetrates inside, as it is the case in flat-roofed perforated plastic greenhouses).
c.  Where to irrigate: Point/line/strip/ or disc source.
d.  How to irrigate: Drip (surface or subsurface) bubbler, micro sprinkler, mist, etc.

## 5.3.2   EVAPOTRANSPIRATION (ET) IN PROTECTED ENVIRONMENT

Evaporation of water requires energy. The availability of energy depends on the microclimate of the protected environment, being the solar radiation the primary source of energy in the ET process. In an unheated greenhouse, the energy used in the ET process can reach 70% of incoming solar radiation [10]. The amount of ground area covered by the crop is the most relevant factor affecting ET. Evaporation (E) from the soil surface is high following irrigation, but decreases rapidly as the soil surface dries. The transpiration (T) will increase with the rise of intercepted radiation (and subsequent increase of ground covered by the crop), while soil evaporation will decrease (as the crop progressively shades the soil surface). Other energy sources (greenhouse heating hot air flow) can increase ET.

Crop evapotranspiration (ETc) or crop water requirements can be related with a reference evapotranspiration (ETo), which is defined as the rate of evapotranspiration from an extended surface of 8–15 cm tall green grass cover of uniform height, actively growing, completely shading the ground, and not short of water [7].

$$ET_c = K_c \times ET_o \qquad (5.1)$$

The crop coefficient (Kc) is the ratio between $ET_c$, and $ET_o$ and depends basically on the crop characteristics, the sowing or planting dates, the development rate of the crop, the length of the cycle, the climatic conditions, and the irrigation frequency, especially at the beginning of the cycle [7].

In greenhouses, the class A pan evaporation method, the radiation (FAO—Food and Agriculture Organization), and Priestley-Taylor methods have been proposed as the more reliable for ET estimation, for periods of several days [5]. The difficulty of an accurate measurement of the wind inside the greenhouse [5] limits the use of the Penman method. The ease of management of the evaporation pan, without sophisticated equipment, is remarkable, but a proper pan placement is necessary. The crop coefficient values for different vegetable crops under greenhouses have been estimated [3, 5, 11, 23]. $ET_c$ can be estimated using USDA class A pan method, as follows:

$$ET_o = K_p \times E_o, \qquad (5.2)$$

$$ET_c = Kc \times ETo = K_c \times K_p \times E_o = K \times E_o, \qquad (5.3)$$

$$K = K_p \times K_c, \qquad (5.4)$$

where, $K_p$ = pan coefficient; and $E_o$ = pan evaporation.

Recent studies show that $K_p$ inside the greenhouse is approximately 1.0 [2, 5, 18] that is higher than open-air values [7]. The crop coefficient evolution and values for different vegetable crops are presented in Table 5.1. Research in the Almeria area confirms that the K values in Table 5.1 show that $K_p$ is around 0.8–0.9, but the quantified values of $K_c$ are higher than those described in the literature [2, 7, 11], being the products of both coefficients ($K_p \times K_c$) similar to those that are indicated in Table 5.1.

**TABLE 5.1**   Crop Coefficient Values for Different Vegetable Crops under Drip Irrigation in a Plastic Green House in Almeria [7].

| Days after sowing or transplanting | Tomato | Capsicum (pepper) | Cucumber | Melon | Watermelon | Beans |
|---|---|---|---|---|---|---|
| 1–15 | 0.25 | 0.20 | 0.25 | 0.20 | 0.20 | 0.25 |
| 16–30 | 0.50 | 0.30 | 0.60 | 0.30 | 0.30 | 0.50 |
| 31–45 | 0.65 | 0.40 | 0.80 | 0.40 | 0.40 | 0.70 |
| 46–60 | 0.90 | 0.55 | 1.00 | 0.55 | 0.50 | 0.90 |
| 61–75 | 1.10 | 0.70 | 1.10 | 0.70 | 0.65 | 1.00 |
| 76–90 | 1.20 | 0.90 | 1.10 | 0.90 | 0.80 | 1.10 |
| 91–105 | 1.20 | 1.10 | 0.90 | 1.00 | 1.00 | 1.00 |
| 106–120 | 1.10 | 1.10 | 0.85 | 1.10 | 1.00 | 0.90 |
| 121–135 | 1.00 | 1.00 | – | 1.10 | 0.90 | – |
| 136–150 | 0.95 | 0.90 | – | 1.00 | – | – |
| 151–165 | 0.85 | 0.70 | – | – | – | – |
| 166–180 | 0.80 | 0.60 | – | – | – | – |
| 181–195 | 0.80 | 0.50 | – | – | – | – |
| 196–210 | 0.80 | 0.50 | – | – | – | – |
| 211–225 | – | 0.60 | – | – | – | – |
| 226–240 | – | 0.70 | – | – | – | – |
| 241–255 | – | 0.80 | – | – | – | – |
| **Total ET$_c$** | **318** | **322** | **156** | **349** | **290** | **146** |

The net irrigation requirements (*IRn*) must replenish the crop evapo-transpirated water (ET$_c$), as rainfall and other components of the water balance are normally unimportant in greenhouses in the Mediterranean area. The gross irrigation requirements (*IR*) must increase the *IRn*, in order to compensate the irrigation efficiency and to leach salts.

$$IRg = \frac{IRn}{Ea(1 - LR)} \tag{5.5}$$

$$Ea = Ks \times Eu \tag{5.6}$$

$$LR = \frac{ECw}{2(maxECe)}, \tag{5.7}$$

where, Ea = irrigation efficiency coefficient (smaller crop root zone to be used by the crop/applied water); Ks is a coefficient (smaller than 1) that expresses the water storage efficiency of the soil (0.9 in sandy soils, 1.0 in clay or loam soils); and Eu is a coefficient (smaller than 1) which reflects the uniformity of water application (a properly designed and well managed drip system should reach the Eu values of 0.85–0.95). This coefficient should be measured for each system regularly [22]; $LR$ = minimum amount of leaching needed to control salts with drip irrigation; $ECw$ = electrical conductivity of the irrigation water (dS/m); $ECe$ = maximum electrical conductivity (dS/m) of the soil saturation extract due to crop withdrawal of soil water to meet its evapotranspiration demand [19, 20].

Typical maximum $ECe$ values are 12.5 in tomato, 10.0 in cucumber, 8.5 in peppers, and 6.5 in bean. Recent research shows that the leaching requirements could be lower than the indicated values [21].

## 5.4   FERTIGATION TECHNOLOGY

The fertigation is an application of fertilizers through the irrigation water. A well-designed fertigation system can reduce cost of fertilizer application considerably and supply nutrients in precise and uniform amounts to the wetted irrigation zone around the tree, where the active feeder roots are concentrated. Applying timely dose of small amounts of nutrients to the trees throughout the growing season has significant advantages over conventional fertilizer practices.

Fertigation saves fertilizer as it permits applying fertilizer in small quantities to supply the nutrient needs of plants. Besides, it is considered eco-friendly as it avoids leaching of fertilizers. Liquid fertilizers are best suited for fertigation. In India, inadequate availability and high cost of liquid fertilizers restrict their uses. Fertigation using granular fertilizers poses several problems namely: the different levels of solubility in water, compatibility among different fertilizers, and filtration of undissolved fertilizers and impurities. Different granular fertilizers have different solubility in water. When the solutions of two or more fertilizers are mixed together, one or more of them may tend to precipitate if the fertilizers are not compatible with each other. Therefore, such fertilizers may be unsuitable for simultaneous application through irrigation water and would have to be used separately. This chapter reports on the various issues of fertigation: advantages and limitations, selection of water soluble fertilizers (granular and liquid), fertigation scheduling in various crops, fertigation systems for efficient fertigation program, and response of plants to fertigation and its economics.

### 5.4.1 IMPORTANCE OF FERTIGATION

The fertigation allows us to apply the nutrients exactly and uniformly only to the wetted root volume, where the active roots are concentrated, which eliminates the over and under feeding of nutrients. Remarkable increase in the application efficiency of the fertilizer saves the significant amount of fertilizers. It allows convenient use of compound and ready-mix nutrient solutions containing also small concentrations of micronutrients which are otherwise very difficult to apply accurately to the soil (Table 5.2). The supply of nutrients can be more carefully regulated and monitored. When fertigation is applied through the drip irrigation, crop foliage can be kept dry thus avoiding leaf burn and delaying the development of plant pathogens.

### 5.4.2 THE FERTILIZATION PROGRAM

a.  **Crop nutrients requirements:** It depends on crop specific needs, yields, methods of growing (open field / protected cultivation), and variety. The crop-specific need may be assessed by mineral analysis of harvested part and vegetative biomass, ratio between N-P-K-Ca-Mg, and percent dry matter. Based on mineral analysis and the yield / plant ratio, one may estimate the crop need for a specific yield and for each ton produced not proportional.
b.  **Soil analysis:** How much N, P, K, Ca, Mg to apply?
c.  **How much nutrients to apply?** It depends on:
    *   **Vegetables:** Fertilization = soil deficit correction + crop nutrients requirement (removed + plant)
    *   **Field crops:** Fertilization = soil deficit correction + removed by yield (harvested)
    *   **Fruit Orchards**: Fertilization = soil deficit correction + removed by yield
    *   **No soil analysis:** The following factors may be used

**TABLE 5.2** Daily Consumption Rate of Nitrogen, Phosphorus, and Potassium (kg·ha⁻¹·day⁻¹) Fertilizers in Different Vegetables under Drip Irrigation after Emergence or Planting [17].

| Days after planting/ emergence | Tomato greenhouse | | | Tomato industry | | | Broccoli | | | Melon | | |
|---|---|---|---|---|---|---|---|---|---|---|---|---|
| | N | P | K | N | P | K | N | P | K | N | P | K |
| 1–10 | 1.00 | 0.10 | 2.00 | 0.10 | 0.02 | 0.10 | 0.02 | 0.00 | 0.01 | 0.15 | 0.03 | 0.10 |
| 11–20 | 1.00 | 0.10 | 4.00 | 0.50 | 0.05 | 0.30 | 0.07 | 0.01 | 0.02 | 0.20 | 0.03 | 0.25 |
| 21–30 | 1.00 | 0.10 | 3.50 | 1.00 | 0.16 | 2.00 | 1.08 | 0.12 | 0.74 | 0.35 | 0.07 | 0.60 |
| 31–40 | 2.50 | 0.20 | 3.50 | 2.80 | 0.19 | 2.30 | 1.22 | 0.13 | 0.91 | 0.90 | 0.18 | 1.45 |
| 41–50 | 2.50 | 0.40 | 5.50 | 4.50 | 0.75 | 8.00 | 1.75 | 0.20 | 1.35 | 1.30 | 0.25 | 3.00 |
| 51–60 | 2.50 | 0.60 | 6.00 | 6.50 | 0.80 | 8.50 | 1.04 | 0.13 | 3.04 | 2.50 | 0.25 | 6.00 |
| 61–70 | 2.50 | 0.30 | 4.00 | 7.50 | 1.80 | 9.00 | 3.02 | 0.36 | 4.34 | 4.30 | 0.35 | 7.00 |
| 71–80 | 2.50 | 0.30 | 6.00 | 3.50 | 0.50 | 4.50 | 3.41 | 0.46 | 3.95 | 2.40 | 0.45 | 8.00 |
| 81–90 | 1.50 | 0.30 | 0.10 | 5.00 | 0.50 | 9.20 | 2.79 | 0.38 | 4.09 | 1.20 | 0.43 | 7.50 |
| 91–100 | 1.50 | 0.10 | 0.10 | 8.00 | 0.89 | 9.00 | 2.09 | 0.32 | 3.13 | 1.00 | 0.27 | 3.50 |
| 101–110 | 1.00 | 0.10 | 0.10 | – | – | – | 0.93 | 0.18 | 2.74 | 0.50 | 0.13 | 1.00 |
| 111–120 | 1.00 | 0.10 | 1.00 | – | – | – | 0.20 | 0.09 | 0.96 | 0.30 | 0.07 | 0.05 |
| 121–130 | 1.50 | 0.20 | 1.00 | – | – | – | 0.18 | 0.09 | 0.48 | – | – | – |
| 131–150 | 1.50 | 0.35 | 1.30 | – | – | – | 0.15 | 0.04 | – | – | – | – |
| 151–180 | 4.00 | 0.50 | 3.80 | – | – | – | – | – | – | – | – | – |
| 181–200 | 2.00 | 0.30 | 3.00 | – | – | – | – | – | – | – | – | – |
| **Total** | **450** | **65** | **710** | **393** | **59** | **520** | **202** | **26** | **165** | **151** | **25** | **385** |
| Date: emergence or planting | 25 Sep** | | | 27 Mar* | | | 30 Aug** | | | 14 Jan | | |
| Harvest | Selective | | | 18 Jul | | | 17 Jan | | | Selective | | |
| Plants/ha | 23,000 | | | 50,000 | | | 33,000 | | | 25,000 | | |
| Soil texture | Sandy | | | Clay | | | Loam | | | Sandy | | |
| Yield (t/ha) | 195 | | | 160 | | | 13 | | | 56 | | |

*Emergence.  **Planting.

**Source:** Scaife, A.; Bar-Yosef, B. Nutrient and Fertilizer Management in Field Grown Vegetables. IPI Bulletin No. 13. International Potash Institute, 1995. https://www.ipipotash.org/udocs/ipi_bulletin_13_fertilizing_for_high_yield_and_quality_vegetables.pdf

$N_{total} \times (1.2$ to $1.3)$: It depends on soil type. Amount of N is lower for heavy soil and higher for light soil.

$P_{total} \times (1.3$ to $2.2)$: It depends on pH and soil type. It increases from light to heavy textured soil and also increases with the increase in pH (5–8).

$K_{exported} \times 1.4$: It depends on soil type. It decreases from low to heavy textured soil.

a. **Method of application:** Surface, sprinkler, and micro irrigation.
b. **Choice of fertilizers:** Granular and liquid.
c. **Timing and quantity to apply:** Growth curve based fertilizer requirement and scheduling.

### 5.4.3  FERTILIZER USE EFFICIENCY (FUE) UNDER DRIP FERTIGATION

Fertigation facilitates the application of water and nutrients directly into the plant root zone, leading to greater efficiencies of application and uptake. This has been substantiated for various crops in studies carried out across the world. Goel [9] reported that FUE of nitrogen is as high as 95% under drip-fertigation compared to 30–50% under soil application (Table 5.3). The results of studies carried out on sugarcane in Maurititus [12] and Australia [6, 16] have indicated increase in nitrogen (N) FUE up to 30%.

**TABLE 5.3**   Fertilizer Use Efficiency (FUE) of Various Nutrients under Fertigation.

| Nutrient | Fertilizer Use Efficiency | | |
|---|---|---|---|
| | Soil application | Drip only | Drip + fertigation |
| Nitrogen | 30–50 | 65 | 95 |
| Phosphorous | 20 | 30 | 45 |
| Potassium | 50 | 60 | 80 |

Timing of fertigation to coincide with periods of demand of the crop (growth–nutrition curve) is a common method of maximizing FUE in many high-value crops with complex phonology and nutrition requirements. It has been reported that splitting nitrogen applications evenly over the first four months of sugarcane crop development led to more efficient and productive use of nitrogen than the growth curve nutrition approach. In another study conducted on sugarcane cultivation under drip-fertigation at PFDC Pantnagar—India showed that sub-surface drip irrigation (SDI) under paired row planting at 75 cm × 75 cm spacing gave the highest cane yield in principal (129.84 tons/ha) as well as ratoon (137.46 tons/ha) crops [14]. This treatment was significantly superior by recording increase of 35.67% in main and 40.15% in ratoon crop over surface irrigated sugarcane planted at 75 cm × 75 cm spacing [14]. Patel and Rajput [13] reported that fertilizer saving of 40% was achieved with fertigation system over conventional practice in okra crop without affecting the yield. They also reported that more than 16% increase in yield under fertigation (16.59–25.21%) over broadcasting method at the 100% level of recommended fertilizer application.

## 5.5 RESULTS AND DISCUSSION

A study (Figs. 5.1 to 5.4) conducted at Department of Irrigation and Drainage Engineering of GBPUA&T, Pantnagar, showed that the water requirement for capsicum varied from 0.090 to 0.8519 lpd/plant for open field conditions, 0.067 to 0.474 lpd/plant for 35% shading conditions, 0.0675 to 0.4605 lpd/plant for 50% shading conditions, and 0.052 to 0.423 lpd/plant for 75% shading condition (Fig. 5.1).

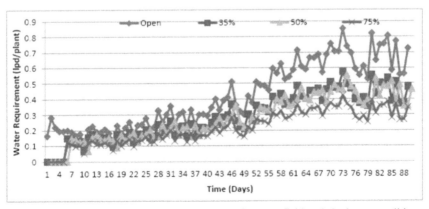

**FIGURE 5.1**  Water requirement of capsicum under open field and shade net conditions. lpd = liters per day.

The water requirement for tomato varied from 0.090 to 0.93 lpd/plant for open field conditions, 0.067 to 0.63 lpd/plant for 35% shading conditions, 0.067 to 0.62 lpd/plant for 50% shading conditions, and 0.059 to 0.46 lpd/plant for 75% shading condition (Fig. 5.2).

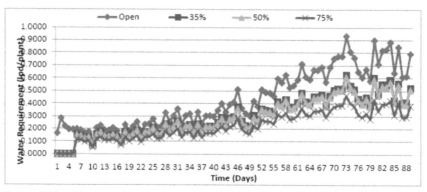

**FIGURE 5.2**   Water requirement of tomato under open field and shade net conditions.

The water requirement for cucumber varied from 0.051 to 0.81 lpd/plant for open field conditions, 0.0385 to 0.545 lpd/plant for 35% shading conditions, 0.038 to 0.529 lpd/plant for 50% shading conditions, and 0.030 to 0.403 lpd/plant for 75% shading condition (Fig. 5.3).

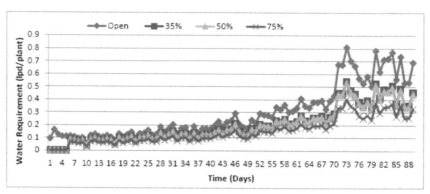

**FIGURE 5.3**   Water requirement of cucumber under open field and shade net conditions.

The water requirement for summer squash varied from 0.0945 to 0.608 lpd/plant for open field conditions, 0.038 to 0.409 lpd/plant for 35% shading conditions, 0.0385 to 0.397 lpd/plant for 50% shading conditions, and 0.030 to 0.302 lpd/plant for 75% shading condition (Fig. 5.4) [15].

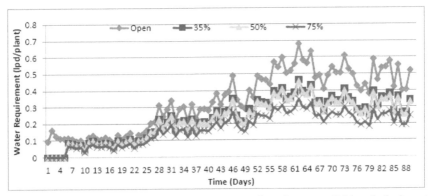

**FIGURE 5.4**   Water requirement of summer squash under open field and shade net conditions.

## 5.5.1   EFFECTS OF DRIP IRRIGATION AND FERTIGATION ON CAPSICUM: A CASE STUDY

A study was conducted at Precision Farming Development Centre (PFDC), Govind Ballabh Pant University of Agriculture and Technology (GBPUA&T), Pantnagar, India [14] on tensiometer based drip irrigation and fertigation scheduling on capsicum under polyhouse. In *Double Span Naturally Ventilated Polyhouse* (DS-NVPH), the water use efficiency (WUE) was observed maximum in the treatment irrigated at 50–70 kPa soil moisture tension and 75% of normal fertilizer dose $(I_3F_1)$ by 281.31% higher compared to the control treatment.

The WUE was increased higher in the treatment by 248.05% compared to the control treatment in WT-NVPH. The water productivity of treatment irrigated at 50–70 kPa soil moisture tension and normal fertilizer doses $(I_3F_2)$ was decreased by −72.43% under DS-NVPH and −71.45% under WT-NVPH in treatment irrigated at 50–70 kPa soil moisture tension and 75% of normal fertilizer doses $(I_3F_1)$ as compared to control treatment (Table 5.4).

**TABLE 5.4** Effects of Tensiometer-Based Irrigation and Fertigation on Capsicum Yield (kg/m²), Water Use Efficiency, and Water Productivity under DS-NVPH (Double Span Naturally Ventilated Polyhouse) and WT-NVPH Conditions.

| Treatment | DS-NVPH | | | WT-NVPH | | |
|---|---|---|---|---|---|---|
| | Yield (kg/m2) | WUE (kg/ m3) | Water productivity (liter/kg) | Yield (kg/m2) | WUE (kg/m3) | Water productivity (liter/kg) |
| $I_1F_1$ | 12.00 | 48.18 | 20.76 | 12.35 | 40.73 | 24.60 |
| $I_1F_2$ | 12.34 | 49.55 | 20.33 | 13.81 | 45.53 | 22.01 |
| $I_2F_1$ | 10.92 | 62.74 | 17.07 | 11.67 | 81.37 | 12.94 |
| $I_2F_2$ | 12.28 | 70.55 | 14.40 | 13.49 | 70.66 | 14.20 |
| $I_3F_1$ | 13.14 | 100.59 | 10.15 | 13.04 | 90.91 | 11.00 |
| $I_3F_2$ | 12.54 | 96.00 | 10.47 | 11.94 | 83.19 | 12.08 |
| C | 8.52 | 26.38 | 37.98 | 9.56 | 26.12 | 38.53 |
| CD (P=0.05) | NS | 20.47 | 4.56 | NS | 17.51 | 4.12 |

Legend: $I_1$ = drip irrigation at 20–30 kPa soil moisture tension; $I_2$ = drip irrigation at 30–50 kPa soil moisture tension; $I_3$ = drip irrigation at 50–70 kPa soil moisture tension; $F_1$ = 75% of recommended fertilizer dose; $F_2$ = 100% of recommended fertilizer dose.

## 5.6 SUMMARY

Application of water and nutrients at the right time, in right amount, in right manner at right place, is the crux of precision water and nutrient management. Micro irrigation provides plants with water through a network of pipe lines at a high frequency but with a low volume of water (drips) applied directly to the root zone in a quantity that approaches consumptive use of the plants, can be combined with fertilizer application, to offer fertigation. Fertigation enables the farmer to meet the specific water and nutrient needs of the crops with great precision, thus minimizing water and nutrients losses. The direct delivery of water and fertilizers through drip irrigation demands the quality of water and use of soluble fertilizers and pumping and injection systems for introducing the fertilizers directly into the irrigation system. Precise irrigation and fertigation allow an accurate and uniform application of water and nutrients to the wetted area, where the active roots are concentrated. The water and nutrients are applied based on the crop need at different growth stages in split dosages. The problem of mobility of non-mobile nutrients is also addressed using fertigation. Planning the irrigation system and nutrient supply to the crops according to their physiological stage of development, and consideration of the soil and climate characteristics, result in high yields and high-quality crops with minimum pollution. In India, more than 4 million ha of land have been brought under pressurized irrigation (sprinkler and micro irrigation). Most of the crops under micro irrigation are horticultural crops. However, field crops such as sugarcane, groundnut, cotton etc. are also being brought under micro irrigation. Efforts are going on to develop economical design of micro-fertigation system for the efficient water and nutrient management in crops like paddy and wheat. The fertilizer applications to these micro irrigated crops are partially through fertigation under open field conditions. However, most of the crops under polyhouse condition are under fertigation. In this chapter, efforts have been made to discuss various issues of irrigation and fertigation scheduling for the precision water and nutrient management for achieving high nutrient use efficiency under protected and open cultivation.

## KEYWORDS

- **fertigation**
- **fertilizers**
- **greenhouse**
- **irrigation scheduling**
- **micro irrigation**

## REFERENCES

1. Ayers, R. S.; Westocot, D. W. *Calidad del agua para la agricultura. Estudio FAO: Riego y Drenaje n°29*; FAO: Roma, 1976.
2. Castilla, N. Contribution to Studies on Covered Crops in Almeria (*Contribucion al estudio de los cultivos enarenados en Almería: Necesidades hídricas y extracción de nutrientes del cultivo de tomate de crecimiento indetermineando en abrigo de polietileno*); Tesis Doctoral. Caja Rural Provincial, Almería, 1986.
3. Castilla, N. Irrigation Programming (*Programacion de l'irrigation goutéà-goutte en serre plastique non chauffée*). *Plasticulture.* **1989,** *82*, 59–63.
4. Castilla, N. Technology Transfer on the Use of Localized Irrigation. In *The Use of Plastics in Agriculture*, Proceedings of XI International Congress, New Delhi, India, 1990; IB, 33–39.

5. Castilla, N.; Fereres, E. The Climate and Water Requirements of Tomatoes in Unheated Plastic Greenhouses. *Agricoltura Mediterranea.* **1990,** *120,* 268–274.

6. Dart, I. K.; Baillie, C. P.; Thorburn, P. J. L. Assessing Nitrogen Application Rates for Subsurface Trickle Irrigated Cane at Bundaberg. *Proc. Aust. Soc. Sugar Cane Technol.* **2000,** *22,* 230–235.

7. Doorembos, J.; Pruitt, W. O. *Water Requirements for Crop* (Las necesidades de agua de los cultivos). Estudio FAO: Riego y Drenaje n" 24. FAO: Roma; 1976.

8. Fereres, E. Drip Irrigation Management. University of California, Berkeley (Calif), 1981.

9. Goel, M. C. Fertigation/Chemigation Scheduling of Liquid Fertilizers for Various Crops. In Proceeding of the Workshop on Micro Irrigation, WTC, IARI, New Delhi; pp 36–59, 2007.

10. Hanan, J. J. The Influence of Greenhouse on Internal Climate with Special Reference to Mediterranean Regions. *Acta Hort.* **1990,** *287,* 23–34.

11. Martinez, A.; Castilla, N. Evapotranspiration for Green House Peppers (*Evapotranspiración del pimiento en invernadero en Almería*). *ITEA.* **1990,** *85,* 57–62.

12. Ng Kee Kwong, K. F.; Paul, J. P.; Devile, J. Drip-fertigation a Means for Reducing Fertilizer Nitrogen to Sugarcane. *Expl Agric.* **1999,** *35,* 31–37.

13. Patel, N.; Rajput, T. B. S. Fertigation—A Technique for Efficient Use of Granular Fertilizer through Drip Irrigation. *IE(I) Journal-(AG).* **2004,** *85,* 50–54.

14. PFDC. *Annual Report 2008–2009 PFDC.* G. B. Pant University of Agriculture and Technlgy, Pantnagar, Uttarakhand, 2009.

15. Rana, S.; Sah, R. Study of the Effect of Protected Environment on Pan Evaporation and Water Requirement of Different Vegetables. U. G. Dissertation, GBPUA&T, Pantnagar, 2010.

16. Ridge, D. R.; Hewson, S. A. Drip Irrigation Management Strategies. *Proc. Aust. Soc. Sugar Cane Technol.* **1995,** *17,* 8–15.

17. Scaife, A.; Bar-Yosef, B. Nutrient and Fertilizer Management in Field Grown Vegetables. *IPI Bulletin No. 13.* International Potash Institute, 1995.

18. Sirjacobs, M. Protected Cultivation of Sweet Peppers in Arid Zone: Evaluation of Water Requirements and Amounts per Watering. *Acta Hort.* **1986,** *191,* 199–207.

19. Stanghellini, C. *Water Balance and Management of Climate in Green House* (Balance hídrico y manejo de microclima en invernadero). IT (Tecnología de invernaderos). Alvarez and Parra (Ed.) Junta de Andalucía. Almería (Spain), 1994; pp 49–62.

20. Stanghellini, C. Evapotranspiration in greenhouses with special reference to Mediterranean conditions. *Acta Hort.* **1993,** *335,* 295–304.

21. Stegman, E.; Musick, J.; Stwart, J. Irrigation Water Management. In *Design and Operation of Farm Irrigation System*; Jensen M. E. Ed., ASAE Monograph 3, Michigan, USA, pp 763–816, 1980.

22. Vermeiren, I.; Jobling, G. A. *Localized Irrigation: Design, Installation, Operation, Evaluation.* FAO Irrigation and Drainage paper 36. Roma, 1980.

23. Veschambre, D.; Vaysse, P. *Practical Guide to Micro Irrigation* (Mémento goutte a goutte: Guide pratique de la micro-irrigation par goîteur et diffuseur); C.T.I.F.L.: Paris, 1980.

# CHAPTER 6

# DRIP IRRIGATION AND FERTIGATION FOR HORTICULTURAL CROPS: SCOPE, PRINCIPLE, BASIC COMPONENTS, AND METHODS

MURTUZA HASAN[*]

*Centre for Protected Cultivation and Technology, Indian Agricultural Research Institute (ICAR), PUSA, New Delhi 110012, India*

[*]*E-mail: mhasan_indo@iari.res.in*

## CONTENTS

## ABSTRACT

This chapter describes technology of drip irrigation that includes components of drip irrigation system, valves and accessories, automation, fertigation methods, and service and maintenance. The chapter also includes basic soil/water/plant relationships, and methods to estimate evapotranspiration.

## 6.1 INTRODUCTION

Water is an essential natural resource for sustaining life and environment. It is imperative that it is utilized with maximum possible efficiency. Agriculture sector is the major user of water resources and the water demand is increasing. The decades of 1960s and 1970s saw the accelerated development of agriculture in India through the intensive use of high-yielding varieties, fertilizers, water, and mechanization. The input-based agricultural planning was successful so that the food production increased several folds during the last three decades. However, the technology based on predominant use of water and fertilizers resulted in a paradoxical situation in which soils in the parts of northern plains turned saline, whereas in some other parts including South India, the water table lowered due to excessive pumping. This situation affected agricultural productivity to a point of stagnation.

In the 1980s, general awareness and consensus emerged on the efficiency and judicious use of water. It was then that the drip irrigation gained popularity with its inherent advantages of water saving and use in problem soils. Various research institutes conducted experiments on drip system and extended the technology to the rural sector. The government also provided liberal support through subsidies to the farmers on procuring and installing drip irrigation systems [2]. Drip fertigation design for protected structures, design of low-pressure drip fertigation system, design of small indigenous low-cost protected structures, and drip fertigation scheduling for various horticultural crops are the important issues related to protected cultivation and drip fertigation [3].

State-wise area covered under drip and sprinkler irrigation systems were approximately 0.690 and 0.450 million ha in 2010–2011 and 2011–2012, respectively (Table 6.1). Rajasthan, Andhra Pradesh, Maharashtra, Karnataka, and Gujarat were leading states in terms of area expansion under drip and sprinkler irrigation. The Government of India (GOI) has been giving subsidy to promote drip and sprinkler irrigation in India. The total subsidy support given to different states of India was approximately Rs. 9970 million and 10,620 million during 2010–2011 and 2011–2012, respectively (in this chapter: Rs. 60.00 = 1.00 US$). Andhra Pradesh, Maharashtra, Rajasthan, and Gujarat were the leading states for receiving financial assistance during 2010–2011 and 2011–2012 to promote drip and sprinkler irrigation in India (Table 6.2).

This chapter presents drip irrigation and fertigation technology for horticultural crops.

**TABLE 6.1**    Statewise Area (ha) under Drip and Sprinkler Irrigation System in India.

| States | 2010–2011 | 2011–2012 (till January, 2012) |
|---|---|---|
| Andhra Pradesh | 122,758 | 91,774 |
| Bihar | 13,485.04 | 14,620.80 |
| Chhattisgarh | 21,830.93 | 16,129 |
| Goa | 119.065 | 34.00 |
| Gujarat | 78,294 | 60,492 |
| Haryana | 9340.2 | 2556.92 |
| Jharkhand | 1217.1 | 0.00 |
| Karnataka | 87,447 | 36,695 |
| Kerala | 2340.01 | 3078.64 |
| Madhya Pradesh | 41,238.24 | 36,544.88 |
| Maharashtra | 118,025.08 | 70,116.86 |
| Odisha | 12,013.96 | 8605.24 |
| Punjab | 4925 | 4026.31 |
| Rajasthan | 147,613 | 87,207 |
| Tamil Nadu | 26,153.16 | 14,228.05 |
| Uttar Pradesh | 3108.63 | 3419.86 |
| West Bengal | 294 | 0 |
| Arunachal Pradesh | 0 | 0 |
| Mizoram | 0 | 0 |
| Meghalaya | 0 | 0 |
| Tripura | 0 | 0 |
| Sikkim | 0 | 0 |
| **Total (India)** | **690,202.42** | **449,528.56** |

**Source:** Lok Sabha Unstarred Question No. 1044, dated March 20, 2012. <www.indiastat. com>

**TABLE 6.2**  Statewise Subsidy Released (million Rs.) by GOI under Drip and Sprinkler Irrigation Systems in India.

| State in India | 2010–2011 | 2011–2012 (up to March 15, 2012) |
|---|---|---|
| Andhra Pradesh | 2400 | 2522.00 |
| Bihar | 000 | 30.00 |
| Chhattisgarh | 101.9 | 200.00 |
| Goa | 2.4 | 2.50 |
| Gujarat | 1200 | 1306.40 |
| Haryana | 136.10 | 169.30 |
| Jammu and Kashmir | 000 | 20.00 |
| Jharkhand | 15.00 | 99.10 |
| Karnataka | 925.40 | 846.40 |
| Kerala | 000 | 20.00 |
| Madhya Pradesh | 796.10 | 886.90 |
| Maharashtra | 2223.70 | 2328.00 |
| Odisha | 81.00 | 82.30 |
| Punjab | 126.10 | 160.00 |
| Rajasthan | 1200.00 | 1309.50 |
| Tamil Nadu | 659.10 | 562.50 |
| Uttar Pradesh | 81.20 | 000 |
| Uttarakhand | 0.00 | 7.50 |
| Arunachal Pradesh | 7.50 | 000 |
| Mizoram | 5.00 | 7.50 |
| Meghalaya | 5.00 | 000 |
| Tripura | 5.00 | 000 |
| Sikkim | 000 | 40.00 |
| Manipur | 000 | 5.00 |
| Nagaland | 000 | 10.00 |
| India | 9970.50 | 10,614.90 |

**Source:** Lok Sabha Unstarred Question No. 1044, dated March 20, 2012. <www.indiastat.com>

## 6.2 DRIP IRRIGATION SYSTEM

Drip irrigation is the best available technology for the judicious use of water for growing horticultural crops on sustainable basis. Drip irrigation is a low labor intensive and highly efficient system of irrigation, which is also amenable to use in difficult situations and problematic soils, even with poor quality water [2]. Irrigation water savings ranging from 36% to 79% can be obtained by adopting a suitable drip irrigation system. Drip irrigation or low-volume irrigation is designed to supply filtered water directly to the root zone of the plant so as to maintain the soil moisture near the field capacity. The field capacity soil moisture level is ideal for efficient growing of vegetable crops. This is due to the fact that at this level the plant gets ideal mixture of water and air for its development. The device that delivers the water to the plant is called dripper.

Water is frequently applied to the soil through emitters placed along a water delivery lateral line placed near the plant row. The principle of drip irrigation is to irrigate the root zone of the plant rather than the soil and getting minimal wetted soil surface. This is the reason for getting very high water application efficiency (90–95%) through drip irrigation. The area between the crop row is not irrigated therefore more area of land can be irrigated with the same amount of water. Thus water saving and production per unit of water is very high in drip irrigation.

Drip irrigation and fertigation technology helps in increasing water and nutritional productivity of horticultural crops. Protected cultivation also helps in increasing water and nutritional productivity of horticultural crops. The possibility of expanding the irrigated areas is becoming very costly, therefore, improving productivity within the existing irrigated area is crucial [1]. The concept of productivity has changed from "crop per unit area" to "crop per unit volume of water." The standard unit of water productivity is $kg/m^3$, whereas nutritional water productivity is expressed in nutritional units/$m^3$ [4]. The crop water productivity is very high for tomato, cucumber, capsicum, and flowers grown inside greenhouse with drip fertigation (Tables 6.3 and 6.4). The crop water productivity of tomato grown inside greenhouse is four times higher than the open field grown tomato. The crop water productivity of green capsicum is higher than colored capsicum grown inside greenhouse.

**TABLE 6.3**   Crop Water Productivity for Greenhouse Vegetables under Drip Fertigation.

| Crop | Greenhouse (GH) | Growing period | Total water use (m3/ha) | Total yield (tons/ha) | Crop water productivity, kg/m3 (g/L) |
|------|-----------------|----------------|-------------------------|------------------------|--------------------------------------|
| Capsicum green | GH | September–May | 2440 | 90 | 37 |
| Capsicum color | GH | September–May | 2440 | 60 | 24.6 |
| Cucumber | GH | August–October | 1550 | 60 | 38.7 |
| Cucumber | GH | February–May | 2010 | 80 | 39.8 |
| Tomato | GH | September–May | 3200 | 250 | 78 |

**TABLE 6.4**   Crop Water Productivity for Greenhouse Flowers.

| Crop | Greenhouse | Growing period | Total water use (m3/ha) | Total yield (stems/ha) | Crop water productivity (stem/m³) |
|------|-----------|----------------|-------------------------|-------------------------|-----------------------------------|
| Chrysanthemum | GH | Yearly | 2800 | 1160,000 | 414 |
| Rose soil | GH | Yearly | 5000 | 2100,000 | 420 |
| Rose soilless | GH | Yearly | 15,000 | 2700,000 | 180 |

## 6.2.1  DRIP IRRIGATION NETWORK: BASIC UNITS

*The pumping unit* takes water from the source and supplies pressurized water to the control head. Pumps used in the drip irrigation system are similar to those used in other irrigation methods and include centrifugal, submersible, and turbine pumps. These pumps can be driven either by an electric motor or an internal combustion engine. An efficiently designed irrigation system has a pumping capacity closely matched to the system demand.

*Control head*: It serves as the irrigation system policeman, regulating flow, pressure, and filtration (Fig. 6.1). It is also the place for chemical injection. Manifold, water meter, and pressure gauge is must for control head. It includes the different types of valves, filters, and hydraulic regulating components.

*Filtration* is the single most critical area in irrigation system. In practical field conditions for intensive growing of vegetable either in greenhouse or in open field, a combination of filtration unit is necessary for proper water treatment. It includes primary and secondary filtration. Several different types of filter can be used to capture and remove contaminants from the irrigation water, such as: gravity filter, mesh screen, sand separator, screen filter, and gravel filter.

**FIGURE 6.1**   Drip fertigation control head.

*Pipe network*: Water is delivered from the control head and filters to the lateral lines in the field through the main and submain pipelines. Main line is mostly of polyvinyl chloride (PVC) material. Submain can be of either PVC or polyethylene. Rigid PVC and polyethylene are typical material used because of the low cost and chemical resistant qualities. In the pressurized irrigation system network, most of the main and submain pipes are buried under the ground and are controlled by various types of control valves.

*Lateral lines and drippers/sprinklers*: The lateral lines supply water to the emitters or drippers from submain. Usually this is placed on the ground. It is made of polyethylene pipes. The diameter varies from 12 to 25 mm. The pressure in the lateral line varies from 1 to 1.2 bars depending upon the lateral length and dripper characteristics. The dripper capacity varies from 2 to 10 L/h at about one bar of working pressure. Some of the common types of dripper used are: in-line dripper, on-line dripper, pressure compensating dripper, and button type dripper.

## 6.2.2   TYPES OF PRESSURIZED IRRIGATION SYSTEMS

A distinction is made between the two principal micro irrigation methods, namely, the spray-jets or micro-sprinkler, and the drip irrigation system (Fig. 6.2). Sprayers and micro-sprinklers spray the water on or near the plant and are designed principally to wet a specific volume of soil around individual trees in an orchard. Drip irrigation, on the other hand, represents a point source of water, and wets a specific volume of soil by direct application of water to the root zone of the plant. The type of drip emitter from the aspect of its discharge and the distribution of the emitters throughout the plot (distances along the drip lateral and between the drip laterals) is dependent on the soil texture and the crop. The drip system is suitable for irrigation of row crops (vegetable and industrial crops) and orchards.

**FIGURE 6.2**   Pressurized drip irrigation system for greenhouse vegetables.

## 6.2.2.1   SPRINKLER IRRIGATION

As stated, the object of sprinkler irrigation is to imitate rainfall. Rain spreads water over large areas in an uniform manner. The sprinklers, which are mounted generally on aluminum or plastic laterals, spray water by a jet, or jets of water (ejected from one or two nozzles), which cause the impact-driven sprinkler to rotate in a circular manner and to spread water over the area according to the radius of through of the jet. The quantity of water accumulated in the immediate vicinity of the sprinkler is generally much greater than that which reaches the soil at the end point of the wetted radius, depending on its particular distribution profile. Hence, in order to obtain high water application uniformity in the field, the sprinklers must be spaced so that the spray from one sprinkler reaches the adjacent sprinkler, i.e., the sprays from adjacent sprinklers must overlap. Spacing of the sprinklers at the correct distances ensures satisfactory uniformity of distribution provided the sprinkler is designed to ensure a regularly shaped distribution pattern and is operated in suitable wind and pressure conditions. Wind adversely affects distribution uniformity. It is therefore, recommended to operate sprinkler irrigation systems in windless conditions or in light wind.

The manufacturers specify the optimal operating pressure for each of their sprinkler models. Operating the sprinkler system at the optimal pressure, and given the other above-mentioned conditions will ensure adequate distribution uniformity. The term application rate (i.e., the amount of water applied to the soil, expressed in mm/h) is a function of the discharge of the sprinklers and their spacing. The infiltration rate of the soil (mm/h) differs for different soils. In order to ensure proper infiltration of the water into the soil and to prevent surface runoff, it is necessary to design the system so that it gives an application rate, which is less than the infiltration rate of the soil. Since sprinklers spread the water in a circular manner, it is recommended to use part-circle sprinklers, which enable adjustment of the area wetted by the sprinkler, so as to minimize loss of water in the area bordering the plot and to avoid interference to vehicles which travel on the field tracks which run along the borders of the irrigated fields. In various stages of field data collection, planning of the irrigation system, and its installation and operation in the field, following considerations will result in high distribution uniformities and irrigation efficiencies:

- Spacing of the sprinklers at distances between them, which will enable attainment of high distribution uniformity;
- Irrigation at an application rate less than the infiltration capacity of the soil;
- Operation of the sprinkler system in windless conditions or light winds only;
- Operation of the sprinkler system according to the pressure recommended by the manufacturer;
- Use of part-circle sprinklers in combination with full-circle sprinklers.
- Portable, i.e., hand-move sprinkler irrigation laterals with adequate overlap can be installed to reduce investment costs.

### 6.2.2.2 SPRAYERS AND MICRO-SPRINKLERS

The micro-sprayer is a device for spreading water through the atmosphere by means of a nozzle and a static spreader platform, whereas the micro-sprinkler spreads the water by means of a rotating, whirling, sprayer device. For this reason, mini-sprinklers with a discharge identical to that of a sprayer can wet wider areas. The micro-sprayer/mini-sprinkler is most suitable for below canopy irrigation of orchard trees. The objective is to wet the soil in a limited area around the tree without wetting all the soil area occupied by each tree. Sprayers have a bridge-like device which enables the static spreader nozzle to be replaced by a rotating device. In this manner, it is possible to irrigate the young orchard by means of micro-sprayers and to switch over to micro-sprinklers when the orchard matures, requiring wetting of a larger soil volume. Micro-sprayer and micro-sprinkler systems can attain maximum irrigation efficiency of 85%.

## 6.3 BASIC COMPONENTS OF DRIP IRRIGATION SYSTEM

### 6.3.1 CONTROL HEAD

It includes pump, filters, fertilizer applicator, water meter, pressure/flow regulating valves, and controller for automation.

## 6.3.2  FILTERS

These remove suspended materials from the water that might clog the drippers. Gravel and graded sand filters are cylindrical tanks that are mainly used for filtering out sand and organic materials after the pumping. Screen and disk filters are the least expensive and most efficient means for filtering water. These are generally used as the second stage filters.

## 6.3.3  FERTILIZER APPLICATORS

These are used to inject fertilizer, systemic insecticides/algaecides, acids, and other liquid materials into the water being supplied through drip system. They are of three types, namely, fertilizer tank, venturi system, and fertilizer pump.

- *Fertilizer tank*: A metallic tank is provided at the head of the drip irrigation system for applying fertilizers in solution along with the irrigation water. The tank is connected to the main irrigation line by means of a bypass line. Some of the irrigation water is diverted from main irrigation line into the tank. This bypass flow is created by the pressure difference between the entry and exit points of the tank. This method is simple in construction and operation. However, the application of fertilizer during the fertigation schedule is not constant. Therefore, it does not permit a precise control over the fertilizer concentration.
- *Venturi system*: It consists of a built-in converging section, throat and diverging section. A suction effect is created at the converging section due to high velocity, which allows the entry of the liquid fertilizer into the system. This system is simple in operation and a fairly uniform fertilizer concentration can be maintained in the irrigation water.
- *Fertilizer pump* is operated either with electricity or with water. It draws fertilizer solution from a tank and pumps it under pressure into the irrigation system. It provides a precision control on the fertilizer application. However, it is expensive and needs skilled operation.

### 6.3.4  PRESSURE/FLOW REGULATORS

These are control valves that are actuated either manually or electrohydraulically to regulate flow and pressure in the drip system.

### 6.3.5  CONTROLLERS

These automatic—mostly microprocessor based—devices are used to provide stop/start signals to pump and valves/regulators. The actuating signal may either be time or volume based. In more advanced technological modes, these gadgets are controlled by soil moisture sensors placed in the plant root zone.

### 6.3.6  PIPE LINES

The water conveyance from the control head to the emitter/dripper is generally categorized in three units as follows:

- *Main pipe*: This is the main carrier of water from the source (after the control head). It is further connected to submains or manifold. These are usually rigid PVC and high-density polyethylene (HDPE) pipes with pressure rating of about 10 kg/cm².
- *Submain*: This is the portion of pipe network between the main pipe line and the laterals. These are also PVC/HDPE/low-density polyethylene (LDPE) pipes with pressure rating of 6–7 kg/cm². Diameters of main and submain pipes are selected based on the water requirements of the farm area.
- *Laterals*: These flexible pipes of HDPE or LLDPE ranging from 10 to 20 mm diameters are the ones which are spread over the field in a specified layout. Designed to carry water at about 3 kg/cm², these pipes are provided with point-source emitters or drippers spaced along it.

### 6.3.7  EMITTER OR DRIPPER

This is a device designed to dissipate the hydraulic pressure and to discharge a small uniform flow of water, drop by drop, at the given place. Different types of emitters have been developed based on the pressure dissipation and flow mechanisms. Emitters are either mounted on the lateral (online type) or fixed inside the lateral (in-line type). The emitters are available in different discharge rate capacities ranging from 1 to 10 L/h. Selection of emitter of given discharge rate depends on the type of crop and soil.

The experimental results conducted on drip irrigation technology have shown 40–50% water saving and increase in yield, besides improvement in the quality of produce. The system cost and net returns from the crop should be carefully analyzed before a decision is made on its installation. Some other factors limiting the large-scale adoption of drip irrigation in India have been identified as under:

- Inadequate knowledge and general awareness about the technology.
- Inadequacy in quality of materials leading to system's failure and short life span.
- Interrupted power supply.
- Damage to rodents.
- Insufficient extension and promotion work by government agencies/ departments.

## 6.4   SOIL–PLANT–WATER–CLIMATE RELATIONSHIPS

Soil–plant–water relationships relate to the properties of soil and plants that affect the movement, retention, and use of water. To understand why and how much irrigation is necessary, one must understand soil–plant–water relations. A proper understanding of these concepts is important to encourage and ensure judicious use of irrigation water and systems.

Water is transported throughout plants almost continuously. There is a constant movement of water from the soil to the roots, from the roots into the various parts of the plant, then into the leaves where it is released into the atmosphere as water vapour through the stomata. This process is called transpiration. Combined with evaporation from the soil and wet plant surfaces, the total water loss to the atmosphere is called *evapotranspiration* (ET).

Soil is a three-phase medium comprising solids, liquid, and air. The fraction of these components varies with soil texture and structure. An active root system requires a delicate balance between the three soil components. However, the balance between the liquid and gas phases is most critical, since it regulates root activity and plant growth process. Soil texture refers to the distribution of the soil particle sizes. The mineral particles of soil have a wide range of sizes classified as sand, silt, and clay.

Soil water affects plant growth directly through its controlling effect on plant water status and indirectly through its effect on aeration, temperature, nutrient transport, uptake, and transformation. The soil and its properties directly affect the availability of water and nutrients to plants. The understanding of soil and its properties in relation to water is helpful in good irrigation design and management. The size, shape, and arrangement of the soil particles and the associated voids (pores) determine the ability of a soil to retain water. Porosity of sandy soils ranges from 30 to 50% compared with 40 to 60% for clay soils.

Water removal from most of the soils will require at least 7 kPa (7 cbars) of tension. The permanent wilting point reaches at around 1500 kPa (15 bars). This is the limit for most plants and beyond this they experience permanent wilting. The pores in sandy soils are generally large and a significant percentage of water drains under the force of gravity in the first few hours after a rain. This water is lost from the root zone to deep percolation. The movement of water from the surface into the soil is called infiltration. The infiltration charactristics of the soil are one of the dominent variables that influence irrigation. Infiltration rate is the soil characteristic determining the maximum rate at which water can enter the soil under specfic conditions. Infiltration rates are generally lower in soils of heavy texture than in light soils. The accumulated infiltration, also called cumulative infilration, is the total quantity of the water that enters the soil in a given time.

### 6.4.1  ESTIMATION OF CROP WATER REQUIREMENT

The estimation of irrigation water requirement is essential for planning cropping as well as irrigation system. Crop water requirement includes the losses due to ET or consumptive use and percolation. The quantity of irrigation water required for a given crop and area can be estimated using the following relationship:

$$W_R = K_C \times K_P\, C_C \times E_V \times A, \tag{6.1}$$

where $W_R$ = monthly/daily irrigation water requirement, liters; $K_C$ = crop factor; $K_p$ = pan evaporation factor (usually 0.8); $C_C$ = canopy factor; it is the ratio of the wetted area to plant area and is taken as 1.0 for closely spaced crops; $E_V$ = monthly/daily pan evaporation, mm; A = area to be irrigated, m²; and I = irrigation interval is the length of time allowable between two successive irrigations.

The crop water requirement under drip irrigation system is different as in traditional irrigation because in drip irrigation system the area between the plant rows are not irrigated and the area between plant to plant is also partially irrigated. Crop water requirement and irrigation interval are the two parameters involved in the irrigation scheduling of any crop. It is expressed either in depth of water (mm or cm) or in amount of water (m³ or L). It depends on soil, plant, climate, and the place of growing.

The pan evaporation method has been successfully used for calculating the crop water requirement of plants on daily basis (in mm/day or cm³/day). The various weather parameters, pan factor, and crop factor are used in the pan evaporation method. The pan evaporation method is simple and practical method of crop water requirement calculation.

## 6.5  FERTIGATION FOR PROTECTED CULTIVATION TECHNOLOGY

In micro irrigation, fertilizers can be applied through the system with the irrigation water directly to the region where most of the plant roots develop. This process is called fertigation and is done with the aid of special fertilizer injectors installed at the head control unit of the system, before the filter. The element most commonly applied is nitrogen. However, application of phosphorous, potassium and other micronutrients are common for different horticultural crops. Fertigation is a necessity in drip irrigation. The main objectives of fertigation are: optimizing yield, minimizing pollution, water saving, fertilizer saving, quality improvement, timely application of fertilizers, and uniform application. The rational for fertigation are following.

- Irrigation and fertilizers are the most important management factors through which farmers control plant development and yield.
- Water and fertilizers have important synergism which is very well used in fertigation.

- Timely application of water and fertilizers can be controlled through fertigation.

### 6.5.1  PRINCIPLE OF FERTIGATION

Fertigation helps to feed the plant at appropriate time, quantity, and location. These three parameters can be controlled through fertigation. The plant yield and the quality depend on all these three factors.

### 6.5.2  ADVANTAGES OF FERTIGATION

- Accurate and uniform application of water and nutrients.
- Application restricted to the wetted area where the active roots are concentrated.
- Amount and concentration of nutrient can be adjusted according to the stage of development and climatic considerations.
- Reduced time fluctuation in nutrient concentrations.
- Crop foliage is kept dry, retarding the development of plant pathogens and avoiding leaf burn.
- Convenient use of ready mixed fertilizers.

### 6.5.3  FACTORS CONTROLLING NUTRIENT UPTAKE

- Water and nutrient distribution in soil under drip fertigation.
- Quantity considerations.
- Intensity considerations (concentration).
- Uptake fluxes: nutrient concentration at root surface.
- Coupling quantity and intensity factors.

### 6.5.4  FACTORS AFFECTING FERTILIZERS' COMPOSITION

- Plant characteristics.
- Soil characteristics.
- Irrigation water quality.
- Growing place.

## 6.5.5 CHEMICALS AND BIOLOGICAL CONSIDERATIONS IN SELECTING FERTILIZERS

- Fertilizers solubility and mixed fertilizers.
- Solution pH and $NH_4/NO_3$ ratio.
- Nutrients mobility and chemistry in soils.
- Salinity of the irrigation water.

## 6.5.6 REQUIREMENTS FOR FERTILIZERS

- Full solubility.
- Quick dissolution.
- High nutrient content.
- Lack of toxic materials.
- Low price.
- Easy availability.

## 6.5.7 FERTIGATION SOLUTION: EC AND PH

Electrical conductivity (EC) and pH are the two important indices of fertigation. They represent the whole quality and characteristics of fertilizers and water. Both vary for different plants and soils. Some important facts related to pH are:

- Alkaline pH may cause precipitation of Ca and Mg carbonates and phosphates.
- High soil pH reduces Zn, P, and Fe availability to plants.
- Ammonia raises the solution pH and urea increases soil pH upon hydrolysis.
- Acids (nitric, phosphoric) may be used to reduce the irrigation solution pH.

## 6.5.8 METHODS OF APPLICATION OF FERTILIZERS

- Fertlizer tank (available in 60, 90, 120 L, etc.)
- Venturi device(head loss/vacuum operated)

- Dosatron (costly and most effective)
- Fertilizer pumps or injectors(hydraulic type)

### 6.5.9  SAFETY DEVICES USED IN FERTIGATION

- An interlock to stop the fertilizer pump.
- A check valve to prevent from the fertilizer tank to the irrigation line following shut down.
- Flow sensor to assure system shut down in case of flow ceases in injection line.
- A bleed valve to relieve the pressure in the injection line when disconnecting.
- A strainer to prevent foreign materials.

### 6.5.10  CHARACTERISTICS OF FERTIGATION

1. High efficiency of fertilizer application.
   - Uniform distribution by irrigation water.
   - Deeper penetration of fertilizer into the soil.
   - Avoiding ammonia volatilizing from soil surface.
2. Easy coordination with specific crops demand.
3. Flexibility in adjusting nutrient ratio.
4. Remote control operation.
   - Allows fertilization in the rainy season when the soil is wet without stepping on it and destroying the structure.
     - Convenience in saving manpower.
     - Low losses in transportation and storage.
     - The system may be used for additional applications.

### 6.5.11  LIMITATIONS OF FERTIGATION SYSTEM

- The system needs water without solid particles that may clog emitters.
- Knowledge of the chemical composition of water is important to avoid precipitation with the added fertilizers. Sometimes pretreatment is necessary.

- The system needs the using of equipment, in which some of them is expensive.
- Not all types of fertilizers are suitable.
- Some fertilizers attack metals mainly of the head control and cause corrosion.
- When proportional fertilizers application is used, nitrate may be leached below the root zone.
- Leaves burning damage occurs, when the fertilizers are applied by sprinklers or micro-sprinklers.

## 6.6 FERTILIZERS

These are chemical compounds in which one atom or more in the formula is a nutrient element. Generally this atom is the ion, or part of an ion that plant adsorbs, apart from the nitrogen in amide group that plant are not able to use it directly. Fertilizers are added to replenish the elements that were used by plants or disappeared by other processes, or will disappear during the plant growth (Tables 6.5 and 6.6).

TABLE 6.5    Fertilizers Generally Used in Fertigation.

| Element | Compound | Formula |
|---|---|---|
| Nitrogen (N) | Urea | $CO(NH_2)_2$ |
| | Ammonium nitrate | $NH_4NO_3$ |
| | Ammonium sulfate | $(NH_4)_2SO_4$ |
| Phosphorus (P) | Phosphoric acid | $H_3PO_4$ |
| | Mono ammonium phosphate | $NH_4H_2PO_4$ |
| | Di ammonium phosphate | $(NH_4)_2HPO_4$ |
| Potassium (K) | Potassium chloride | KCL |
| | Potassium nitrate | $KNO_3$ |
| | Mono potassium phosphate | $KH_2PO_4$ |

TABLE 6.6    The Ionic Form of Nutrition Elements Adsorbed by Plants.

| Cations | Anions |
|---|---|
| Ammonium $NH_4^+$ | Nitrate $NO_3^-$ |
| Potassium $K^+$ | Monophosphate $H_2PO_4^-$ |
| Calcium $Ca^{++}$ | Di-phosphate $HPO_4^{--}$ |

**TABLE 6.6**   *(Continued)*

| Cations | Anions |
|---------|--------|
| Magnesium $Mg^{++}$ | Sulfate $SO_4^{--}$ |
| Iron $Fe^{+++}$ | Molybdate $MoO_4^{--}$ |
| Iron $Fe^{++}$ | Borate $B_4O_7^{--}$ |
| Manganese $Mn^{++}$ | Copper $Cu^{++}$ |
| Zinc $Zn^{++}$ | |

## 6.7   LOW-PRESSURE DRIP IRRIGATION TECHNOLOGY

It is a new innovation in pressurized irrigation technology. Pressurized irrigation technology like drip irrigation and sprinkler irrigation has many advantages mainly in the form of water and fertilizer saving, increase in the crop production, removal of weeds, etc. As the name suggests, these technologies need pressure and energy mainly in the form of electrical energy for their working operation. Thus, the pressurized irrigation technologies were totally dependent on supply of electricity. It was the major bottleneck of these technologies.

A new innovation has been made in the field to run the drip irrigation system in the small land holding with gravitational energy rather than commonly used electrical energy. It has been named as low-pressure drip irrigation technology or gravity-fed drip irrigation. In this system, gravitational energy is used by placing the water supply tank at the height of minimum 1.5 m. The platform of locally available materials like brick, stone, wood, plank is made of minimum 1.5-m height to place the water tank of 500–1000 L over it. Normally, 1000-L tank is sufficient to irrigate 1000 m² area of horticultural crops.

The lateral or bed length used in this system is not more than 20 m (Fig. 6.3). The lateral pipe of 12–16 mm fitted with dripper of discharge 1 Lph is commonly used in this system. The hydraulics of low-pressure drip irrigation system has been studied through different experiments, which suggest the optimum use of water and nutrients in this particular model. The major advantage of this system is the simplification in the use of fertigation. In the pressurized irrigation system, there is a need of extra pump, venturi, or tank for the supply of fertilizers. This requires additional money and energy to be used in the system. In low-pressure drip irrigation technology, same water supply irrigation tank is used for supply of fertilizers and other micronutrients.

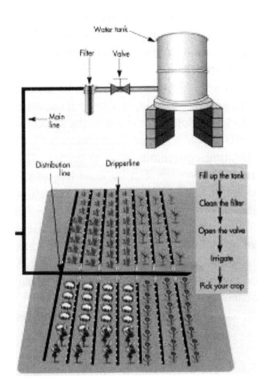

**FIGURE 6.3** Layout of low-pressure drip irrigation technology.

Normally the commonly used fertilizers and micronutrients are directly used in the irrigation tank and supplied to the crops. Thus, we can see that low-pressure drip irrigation technology has simplified and economized the pressurized drip irrigation technology. This system is now particularly suitable for Indian villages and Indian farmers. Low-pressure drip irrigation technology is getting into the villages due to its advantage over pressurized drip irrigation technology for the use of the system without electricity as many Indian villages have erratic electricity supply. The system is very popular in the farming community as it is technically simple and easy to use in the field. The system can be installed, used, maintained, and replaced by one small family. Therefore, it is sometimes also known as family drip irrigation system. Land holding area is decreasing in India due to increasing population and there is a shift toward use of protected cultivation and use of horticultural crops. In all these scenarios, low-pressured drip irrigation technology is extremely suitable and beneficial for Indian farmers. Low-pressure drip irrigation technology is particularly suitable for protected horticulture and greenhouse farming, where the land holding is small and there is a precise need of water and fertilizers.

## 6.8  TECHNICAL DETAILS OF LOW-PRESSURE DRIP IRRIGATION TECHNOLOGY

Low-pressure drip irrigation system is particularly suitable for protected horticulture where the crop grown area is relatively small and precise fertigation is required to get the quality products. This system requires the hydraulic head of about 2 m for irrigating an area of about 500 m². It requires comparatively less pressure than normal pressurized irrigation system and is suitable for small land holdings, greenhouses and nursery. It has high irrigation efficiency (greater than 90%) and is easy to install, operate, and maintain. The low-pressure drip irrigation system overcame the two major limitations of the pressurized irrigation system viz, continuous requirement of energy and high initial cost. The high cost and energy requirement in the conventional drip irrigation system deters its adoption on a large scale. Consequently, only about 1% of the total land irrigated in the world is done by drip irrigation. Low-pressure drip irrigation system requires comparatively less pressure than normal pressurized irrigation system and is suitable for small land holdings, greenhouses, and nursery.

This system does not require regular supply of electrical energy and the hydraulic head of about 2 m is sufficient for irrigating an area of about 500 m². This system has high irrigation efficiency (greater than 90%) and is easy to install, operate, and maintain.

### 6.8.1   COMPONENTS OF GRAVITY-FED DRIP IRRIGATION SYSTEM (FIG. 6.3)

- 500-L water tank.
- 30-mm diameter HDPE (high-density polyethylene) main line pipe of about 30 m.
- One inch 120-mesh/130 micron disc/screen filter.
- Tank connectors.
- 500-m length lateral LDPE pipe of 12–16-mm dia fitted with in-line dripper of 1 Lph discharge capacity.

## 6.9   AUTOMATION OF DRIP IRRIGATION SYSTEM

Automation of irrigation system in protected cultivation mainly in the form of greenhouse, net house, and nursery is practiced to make the irrigation operation precise, perfect, and efficient. The automated irrigation system also minimizes the number of personnel involved in irrigation operation and makes the irrigation system relatively maintenance free. The introduction of automation into irrigation systems has increased application efficiencies and drastically reduced labor requirements. The automation of irrigation system is mainly of two types: conditional and nonconditional.

The conditional automated irrigation system depends on various types of sensors placed in the irrigated area. The opening and closing of the valves to operate the irrigation system depends on the data of the sensors passed on the programs installed in the irrigation computer. The examples of these sensors are humidity, temperature, tensiometer sensors. Conditional automation of irrigation system is very efficient but it requires smooth operation of sensors and thereby more maintenance. It results in precise irrigation as and when required by the plant to meet its crop water requirement. Consequently the overall irrigation efficiency is very high in this case.

The nonconditional automated irrigation system depends on the numeric data fitted manually in the irrigation computer to different irrigation programs. These programs operate the opening and closing of the valves attached to different crops for irrigation. The different irrigation programs need to be updated manually to suit the crop water requirement. This system requires experienced irrigation personnel to update different irrigation programs. In this case the data of the various types of sensors are not directly connected to irrigation programs but it serves as a guide to manually update the irrigation programs installed in the controller or computer. The maintenance of sensors is also necessary to operate this system efficiently.

### 6.9.1 BASIC ELEMENTS OF AN AUTOMATIC IRRIGATION SYSTEM

- *Metering device*: A water meter with electrical output is installed in the irrigation system; information relating to accumulating flows can be monitored and relayed by cables to the control unit. Fertilizer meter is also installed to transmit the information concerning cumulative volumes of a fertilizer solution which is pumped from a tank and injected into the system.
- *Electronic control unit*: The control unit receives the necessary feedback from the metering or measuring devices, and processes the data in order to relay the proper commands to the appropriate valves.
- *Solenoid valves*: It converts the electrical commands received from the control unit into hydraulic commands, which causes the valves to open or close.
- *Hydraulic valves*: These control the flow of water into single laterals or sets of laterals operating simultaneously.

Applying water directly near to the root zone of plant by switching on the pump is itself the first step toward the automation of irrigation system. The complete automation of micro irrigation system is a relatively new concept in rural areas of India. The inclination toward automation of micro irrigation is gaining momentum due to following reasons:

- Automation eliminates manual operation to open or close valves, especially in intensive irrigation process.
- Possibility to change irrigation and fertigation frequency and also to optimize these processes.
- Adoption of advanced crop systems and new technologies, especially new crop systems that are complex and difficult to operate manually.
- Use of water from different sources and increased water and fertilizer use efficiency.
- System can be operated at night to save water from evaporation loss and thus the day time can be utilized for other agricultural activities.
- Pump starts and stops exactly when required, thus optimizing energy requirements.

## 6.10   DESIGN OF DRIP IRRIGATION SYSTEM

A complete drip irrigation system, consists of a head control unit, main, and submain pipelines, hydrants, manifolds, and lateral lines with drippers.

*Control station* (head control unit): Its features and equipment depend on the systems requirement. Usually, it consists of the shut off, air and check valves, filtering unit, fertilizer injector, and other small accessories.

*Main and submain pipelines*: They are made of PVC and are mostly buried in the ground.

*Hydrants*: Fitted on the mains or the submains and equipped with 2–3 shut off valves, they are capable of delivering all or part of the piped water flow to the manifold feeder lines.

*Dripper lateral*: They are made of 12–20-mm LDPE PN 3–4 bars. They are equipped with in-line or online drippers.

*Dripper*: The system pressure ranges from 2 to 3 bars. Operating pressure of dripper is usually one bar. Dripper discharge varies from 1 to 10 Lph. The wetting pattern of water in soil from the drip irrigation tape must reach plant roots. Emitter spacing depends on the crop root system and soil properties. Seedling plants such as onions have relatively small root systems, especially early in the season. Design must take into account the effect of the land's contour on pressure and flow requirements. Plan for water distribution uniformity is by carefully considering the tape, irrigation length, topography, and the need for periodic flushing of the tape.

It should include vacuum relief valves into the system. Consider power and water source limitations. Finally, be sure to include both injectors for chemigation and flow meters to confirm system performance.

*Filters and pumps*: Sand media filters have been used extensively for micro irrigation systems. Screen filters and disk filters are common as alternatives or for use in combination with sand media filters. Sand media filters provide filtration to 200 mesh, which is necessary to clean surface water and water from open canals for drip irrigation. These water sources pick up a lot of fine grit and organic material, which must be removed before the water passes through the drip tape emitters. Sand media filters are designed to be self-cleaning through a "backflush" mechanism. This mechanism detects the drop in pressure due to the accumulation of filtered particles. It then flushes water back through the sand to dispose of clay, silt, and organic particles. Sand used for filters should be between sizes 16 and 20 to prevent excess back flushing. To assure enough clean water for back flushing, several smaller sand media filters are more appropriate than a single large sand media. In addition to a sand media filter, a screen filter can be used as a prefilter to remove larger organic debris before it reaches the sand media filter, or as a secondary filter before the irrigation water enters the drip tube.

For best results, filters should remove particles four times smaller than the emitter opening, as particles may clump together and clog emitters. Screen filters can act as a safe guard if the main filters fail, or may act as the main filter if a sufficiently clear underground water source is used.

### 6.10.1 CONSIDERATIONS TO DESIGN AN EFFECTIVE DRIP IRRIGATION SYSTEM

- Work out the number of connectors needed when planning your drip irrigation system.
- Plants in sunny areas usually require more water due to higher evaporation rates. Plants in shaded areas will require less water due to lower evaporation rates.
- Note slopes and soil types to help work out the watering requirements for different areas of your garden. For example, gardens with heavy clay soil may need more water pressure.
- Select drip emitters according to your plants' watering requirements.

- Consider where you would need joints and connectors.
- Rain switches and soil moisture sensors are highly recommended, especially in areas with high rainfall.
- Lay the piping above ground before digging. A 10-cm deep trench should be adequate, although sandy soil may require a slightly deeper trench to hold the piping in place.
- To make it easier to connect joints, heat the end of your piping to soften it and make it more flexible.
- Make sure drip emitters are installed above ground so that they do not become clogged by dirt.

## 6.11 MAINTENANCE OF DRIP IRRIGATION SYSTEM

The usual life of drip irrigation system is 7–10 years. It can be achieved only by regular maintenance of different components of drip irrigation system. The drip irrigation system can become nonoperational or ineffective if all its components are not maintained regularly. Most of the maintenance jobs can be done very easily without any extra expenditure. Some of key components of maintenance of drip irrigation system are described below:

- *Filters* should be checked and flushed on a regular basis. When the system is new, check after a few days to see what kind of dirt has been caught in the filter. If there is really nothing go ahead and check again in a month or two. Just make sure to take out the screen and rinse it out under clean water when it needs it. This will keep the system running smoothly for a long time. Filtration is one of the most important part of a drip system, treat it as such!
- *Drippers* really do not need any kind of regular maintenance. There are models which come apart and can be cleaned. To do this take them apart and rinse under clean water. For stubborn dirt use an old toothbrush. Sometimes drippers will just get clogged up with hard water deposits or dirt and it is almost always easier to replace them. Most drippers will give you a few years of good service before any kind of problems arise. The life of drippers varies from 6 to 8 years.
- *Controllers* need very little checkup and checking the condition of the batteries is necessary. We suggest replacing the batteries every year just to make sure they will always be good. Controllers require

uninterrupted regular power supply. Some arrangements should be there to regulate the supply.

- *Valves* can be affected by debris in the water which could collect over time. This can be fixed by disassembling the valve and cleaning it. All the valves can be taken apart and cleaned. The instruction manual will generally indicate how to dismantle and put—together a particular type of valve.

- Place a *water flow meter* between the solenoid valve and each zone and record it daily. This provides a clear indication of how much water is applied to each zone. Records of water flow can be used to detect deviations from the standard flow of the system, which may be caused by leaks or by clogged lines. The actual amount of water applied recorded on the meter can be compared with the estimated crop water use (crop ET) to help assure efficient water management.

- *Leaks* can occur unexpectedly as a result of damage by insects, animals, or farming tools. Systematically monitor the lines for physical damage. It is important to fix holes as soon as possible to prevent uneven irrigation.

- *Chlorine clears clogged emitters*: If the rate of water flow progressively declines during the season, the tubes or tape may be slowly plugging, resulting in severe damage to the crop. In addition to maintaining the filtering stations, regular flushing of the drip tube and application of chlorine through the drip tube will help minimize clogs. Once a month, flush the drip lines by opening the far ends of a portion of the tubes at a time and allowing the higher velocity water to rush out the sediment.

- *Plan for seed emergence*: The drip tape must be close enough to the surface to germinate the seed if necessary, or a portable sprinkler system should be available. For example, a tape tube 4–5 inches deep has successfully germinated onion seeds in silt loam soil. Tape at 12 inches failed to uniformly germinate onions.

- *Timing and rates*: The total irrigation water requirements for crops grown with a drip system is greatly reduced compared with a surface flood system because water can be applied much more efficiently with drip irrigation. For example, with furrow irrigation, typically at least 4-acre-feet/acre/year of water is applied to onion fields in the Treasure Valley of Eastern Oregon and Southwestern Idaho.

Depending on the year, summer rainfall, and the soil, 14–32 acre-inches/acre of water has been needed to raise onions under drip irrigation in the Treasure Valley. Applying more water than plants need will negate most of drip irrigation's benefits. The soil will be excessively wet, promoting disease, weed growth, and nitrate leaching.

### 6.11.1  STANDARD MAINTENANCE PRACTICES FOR DRIP FERTIGATION

- Add chlorine or other chemicals to the drip line periodically to kill bacteria and algae. Acid might also be needed to dissolve calcium carbonates.
- Filters must be managed and changed as needed. Even with filtration, however, drip tape must be flushed regularly. The frequency of flushing depends on the amount and kinds of sedimentation in the tape.
- Root intrusion needs to be controlled for some crops. Rodents must be controlled, especially where drip tape is buried

### 6.12  MAINTENANCE OF DRIP FERTIGATION SYSTEM FOR PROTECTED CULTIVATION

- The most commonly occurring problem in drip irrigation system is the clogging of drippers. Once clogged, it is very difficult to declog especially the in-line drippers. Clogging can be avoided by keeping a regular maintenance schedule of filters.
- Upkeep of components is very critical to make the drip operation successful. Gravel/sand filters must be washed through backflow to remove the sedimentation inside. Disk of screen filers should be washed every alternate day. The filtering elements of both these type of filters are washable. Rubber or seals should be replaced properly so that no leakage occurs.
- All emitters should be periodically checked for proper functioning. Any leakage in the pipes; fittings should be checked immediately.
- Fine inorganic particles usually settle at the ends of the submain manifold and the laterals where flow velocities are slow. Periodic flushing should be done by removing end plugs of laterals and

flush out the fine particles. A velocity of 30 cm/s is necessary to adequately flush the fine particles from lateral tubing.

- Generally online type of emitters can be disassembled and cleaned manually. In-line type can be flushed to eject loose deposits. Carbonate deposits can removed using solutions of 0.5–1.0% HCL acid ejected at the submain lateral connection to give a contact time of 5–15 min in the emitter. This treatment may not be effective in completely clogged emitters.
- Emitters discharge rate should be periodically checked for uniformity.

## 6.13  SUMMARY

Drip Irrigation and fertigation technology has emerged as potential water-saving and yield-enhancing technology mainly for horticultural crops. It helps in increasing crop water productivity toward sustainable level for different horticultural crops grown in diverse agroclimatic zones. The principle of drip irrigation is to irrigate the root zone of the plant rather than the soil and getting minimal wetted soil surface. The high initial investment and being technology sensitive are the two major bottlenecks for the wide scale adoption of this technology. Nevertheless, it has been expanding throughout the world. The two major types of drip fertigation system are low-pressure and pressurized drip fertigation technology that have been discussed in this chapter. The drip irrigation consists of: control units, filtration unit, fertigation unit, distribution unit, emitters, controls, and automation.

## KEYWORDS

- **automation**
- **efficiency**
- **fertigation**
- **greenhouse**
- **sprinkler irrigation**

## REFERENCES

1. Carruthers, I. M.; Rosegrant. W.; Seckler. D. Irrigation and Food Security in the 21st Century. *Irrig. Drain. Syst*. **1997,** *11*, 83.
2. Goyal, M. R. *Management of Drip/Trickle or Micro Irrigation*; Apple Academic Press Inc.: Oakville, Canada, 2012.
3. Hasan, M. Protected Cultivation and Drip Fertigation Technology for Sustainable Food Production. In Souvenir of The Second International Conference on Bioresource and Stress Management, Hyderabad, India, Jan 7–10, 2015, pp 19–24.
4. Renault, D.; Wallender, W. W. Nutritional Water Productivity and Diets. *Agric. Water Manage*. **2000,** *45*, 275–296.

# PART III
# Design of Drip Irrigation Systems

## CHAPTER 7

# DESIGN OF MICRO IRRIGATION SYSTEM: SLOPING AND TERRACED LAND

P. K. SINGH*, K. K. SINGH, R. SINGH, and H. S. CHAUHAN

*Irrigation and Drainage Engineering Department, College of Technology, G. B. Pant University of Agriculture and Technology, Pantnagar 263145, Uttarakhand, India*

*Corresponding author. E-mail: singhpk67@gmail.com*

## CONTENTS

## ABSTRACT

Design of micro irrigation system must be in accordance with the crop demand, soil texture, and agroclimatic conditions of the location, for achieving the maximum productivity of quality produce, including the conservation of precious water and land resources. Design of micro irrigation system depends on several parameters, including topography, soil texture and crop to be irrigated, weather conditions, technology, and financial resources. Different criteria are available for design of the micro irrigation system for widely spaced row crops such as orchard and vegetables for supplying water to the individual plants with the help of a single or a set of drippers based on the rooting pattern and canopy area of the plants. The design of main/sub-main for the slopping and terraced land having water source at the top or bottom of field were also considered in the design.

## 7.1  INTRODUCTION

There is tremendous variation in topography of land, land shape (trapezoidal or rectangular), and water source availability (at top or bottom of the field) for growing agricultural and horticultural crops. The land has varying degree of slopes, soil depth, and fertility level. To have irrigated land for fruits, vegetables, medicinal and aromatic plants, important cash crops and newly established plantation crops, and also to improve productivity of existing old orchards, it is necessary to develop irrigation water resources and methods for application of irrigation water in an efficient manner. The development of irrigation water resources in the hills is possible through taping/harnessing of rainwater and runoff, water flow in streams, *nalas*, and springs of different dimensions, and taping of natural existing springs, and several low-discharge natural springs. The natural spring constitutes promising water resources in hills. The discharge of these springs varies with the season and rainfall conditions. These natural springs have good potential of water for domestic and irrigation purposes of agricultural crops. To increase the irrigation potential in hills, it is utmost necessary that existing water resources in the uplands (such as springs, streams, and surface/subsurface runoff) should be utilized [2]. Storing water in tanks, the low-discharge springs/streams can be used to provide assured irrigation round the year in hilly/sloping land system. Whereas, the surface runoff stored in the water harvesting tanks can be used for supplemental irrigation during summer or nonrainy season. Modern irrigation technique such as micro irrigation would have special advantage in the sense that the available gravity head can be used without requiring an additional pumping unit and energy. The soils of the hill lands are generally characterized by high intake rate. Such types of soil require frequent and small depth of water application for better crop growth and high water use efficiency (WUE). Micro irrigation system in hills can provide a system through which a smaller depth of water is applied more frequently and in a larger area.

Design of micro irrigation system must be in accordance with the crop demand, soil texture, and agroclimatic conditions of the location, for achieving the maximum productivity of quality produce, including the conservation of precious water and land resources. Design of micro irrigation system depends on several parameters, including topography, soil texture and crop to be irrigated, weather conditions, technology, and financial resources. Different criteria are available for design of the micro irrigation system for widely spaced row crops such as orchard and vegetables for supplying water to the individual plants with the help of a single or a set of drippers based on the rooting pattern and canopy area of the plants. In this situation, there is no need to apply water to the entire land area and the laterals are generally placed along the plant rows. In case of closely spaced crops, the entire land area needs to be irrigated and the micro irrigation system needs to be designed on the basis to meet the water requirement of the total cultivated area. Relationship between emitter discharge, operation time, horizontal and vertical movement of soil moisture under micro irrigation provides superior criteria for designing efficient and economic micro irrigation system for closely spaced crops. In general, pumping unit is an essential component of micro irrigation system for creating desired water pressure head required for operating emitters.

Main/sub-main lines are important components of micro irrigation system. It is a multi-outlet pipe lines which supplies water to laterals (single/pairs) and ultimately to the plants through emitters or drippers. The discharge and the spacing of sub-mains depend on the number and flow rate of the laterals. The design of micro irrigation main or sub-main should be based on the balance between allowable variation in head, friction head loss, and the elevation difference in sub-main [1]. In the design of sub-main for sloping/terraced land, the head loss due to bends in sub-main is also considered. In the steep sloping lands, the difference in elevation between the inlet and end points of sub-main is sufficiently high and hence the pressure compensating emitters or drippers can be used under these conditions to compensate head due to higher elevation differences. The design of main/sub-main for the slopping and terraced land having water source at the top or bottom of field were also considered in the design.

## 7.2   DESIGN OF MAIN/SUB-MAIN LINES FOR MICRO IRRIGATION

Head loss due to friction in smooth pipes can be estimated by most commonly used Hazen–Williams Equation (7.1) and the head loss due to pipe fittings can be computed by Equation (7.2):

$$h_f = K \frac{(Q/C)^{1.852}}{D^{4.872}} \times \frac{L}{100},$$   (7.1)

$$h_{fb} = K_b K_r \frac{Q^2}{D^4},$$   (7.2)

where, $Q$ is the discharge rate in pipe (L/s); $C$ is friction coefficient; $D$ is inner diameter of pipe (mm); $L$ is length of pipe (m); $K$ and $K_b$ are constants ($K = 1.22 \times 10^{12}$, $K_b = 8.26 \times 10^4$ for metric units); $K_r$ is resistance coefficient for the fittings; $h_f$ is the head loss due to pipe friction (m) and $h_{fb}$ is the head loss due to pipe fittings; considering a sloping terraced land in which $S$ is the ground slope and $h_r$ and $W_t$ are the riser height and terrace width of bench terraces made on that slope, respectively (Fig. 7.1). It is considered that $L_{t1}$, $L_{t2}$, ..., and $L_{tm}$ are the length of first, second, ..., and nth terrace, respectively.

Thus, the number of plants at the ith terrace is $Ni = L_{ti}/S_p$ and the discharge of ith lateral will be $q_r = N_i \times N_p \times q_a$, where: $N_p$ is the number of emitter/drippers per plant; $q_a$ is the average emitter/dripper discharge (lph) and $q_1$ is the discharge rate of lateral on ith terrace (lph). The sub-main flow rate can be calculated as: $Q_m = q_1 + q_2 + ... + q_1 + ... + q_n$. The flow rate in the sub-main on ith terrace will be less than the flow rate on (i − 1) th terrace. Thus, if the flow rates of the sub-main on the first, second, ..., and nth terrace are $Q_1$, $Q_2$, ..., and $Q_n$, respectively, then the head loss due to friction and bends of pipe can be calculated using eqs (7.1) and (7.2), respectively.

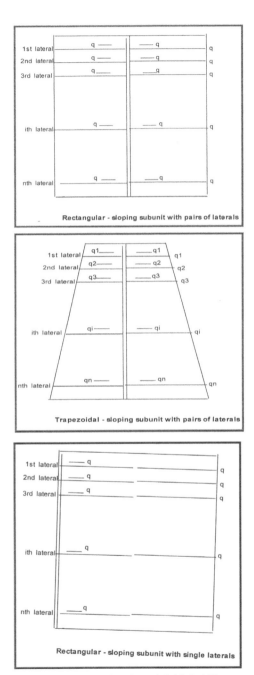

**FIGURE 7.1** Trapezoidal and rectangular-shaped fields in hilly area.

### 7.2.1 FIRST TERRACE

From Figure 7.1, it is clear that the sub-main flow rate in the first terrace $Q_1 = Q_m$ and the length of sub-main on first lateral $L_1$ and number of bends up to first terrace is equal to 2. Thus, head loss up to first terrace, $h_1$ = head loss due to bend + head loss due to pipe friction. Inserting these values in eqs (7.1) and (7.2), the head loss up to first terrace is given as follows:

$$h_1 = 2\frac{K_b K_r Q_1^2}{D^4} + K\frac{(Q_1 / C)^{1.852}}{D^{4.87}} \times \frac{L_1}{100}. \tag{7.3}$$

### 7.2.2 SECOND TERRACE

Discharge after first terrace, $Q_2 = Q_m - q_1$. Length of sub-main in between first and second terrace, $L_2 = h_r + W_t$. Number of pipe bends = 2. Head loss between first and second terrace is given as follows:

$$h_2 = 2\frac{K_b K_r Q_1^2}{D^4} + K\frac{(Q_1 / C)^{1.852}}{D^{4.87}} \times \frac{L_2}{100}. \tag{7.4}$$

### 7.2.3 THIRD TERRACE

Discharge after second terrace, $Q_3 = Q_m - (q_1 + q_2)$. Length of sub-main in between second third terrace, $L_3 = h_r + W_t$. Number of bends in third terrace = 2. Head loss between second and third terrace is given as follows:

$$h_3 = 2\frac{K_b K_r Q_1^2}{D^4} + K\frac{(Q_1 / C)^{1.852}}{D^{4.87}} \times \frac{L_3}{100}. \tag{7.5}$$

### ith Terrace

Discharge after $(i-1)$th terrace, $Q_i = Q_m - (q_1 + q_2 + \dots + q_{i-1})$. Length of sub-main in-between $(i-1)$th and ith terrace, $L_i = h_r + W_t$. Head loss in ith terrace can be given as follows:

$$h_i = 2\frac{K_b K_r Q_1^2}{D^4} + K\frac{(Q_1/C)^{1.852}}{D^{4.87}} \times \frac{L_i}{100}. \qquad (7.6)$$

### nth Terrace

Discharge after (n − 1)th terrace $Q_n = Q_m - (q_1 + q_2 + \ldots + q_{n-1})$. Length of sub-main in between (n − 1)th and nth terrace, $L_n = h_r + W_t$. Number of bends in nth terrace = 2. The total head loss in nth terrace can be given as follows:

$$h_n = 2\frac{K_b K_r Q_1^2}{D^4} + K\frac{(Q_1/C)^{1.852}}{D^{4.87}} \times \frac{L_n}{100}. \qquad (7.7)$$

## 7.2.4   TOTAL HEAD LOSS FOR ALL TERRACES

Total head loss due to friction and pipe bends (from first terrace to nth terrace) can be obtained by adding the head loss in individual terraces. Thus, adding the Equations (7.3) to (7.7) we have:

$$hfb = [2\frac{K_b K_r Q_1^2}{D^4} + K\frac{(Q_1/C)^{1.852}}{D^{4.87}} \times \frac{L_1}{100}] + [2\frac{K_b K_r Q_1^2}{D^4} + K\frac{(Q_1/C)^{1.852}}{D^{4.87}} \times \frac{L_2}{100}]$$

$$+ [2\frac{K_b K_r Q_1^2}{D^4} + K\frac{(Q_1/C)^{1.852}}{D^{4.87}} \times \frac{L_3}{100}] + \ldots + [2\frac{K_b K_r Q_1^2}{D^4} + K\frac{(Q_1/C)^{1.852}}{D^{4.87}} \times \frac{L_i}{100}]$$

$$+ [2\frac{K_b K_r Q_1^2}{D^4} + K\frac{(Q_1/C)^{1.852}}{D^{4.87}} \times \frac{L_n}{100}.] \qquad (7.8)$$

On rearranging the Equation (7.8), we have:

$$hfb = 2\frac{K_b K_r}{D^4}\left[Q_1^2 + Q_2^2 + Q_3^2 + \ldots Q_i^2 + \ldots Q_n^2\right] +$$

$$\frac{K}{D^{4.87}}\left[\left\{\frac{Q_1}{C}\right\}^{1.852} \times \frac{L_1}{100} + \left\{\frac{Q_2}{C}\right\}^{1.852} \times \frac{L_2}{100} + \left\{\frac{Q_3}{C}\right\}^{1.852} \times \frac{L_3}{100} + \ldots \left\{\frac{Q_i}{C}\right\}^{1.852} \times \frac{L_i}{100} + \ldots \left\{\frac{Q_n}{C}\right\}^{1.852} \times \frac{L_n}{100}\right]. \qquad (7.9)$$

Simplifying Equation (7.9), we get:

$$h_{fb} = 2\frac{K_b K_r}{D^4}\sum_{i=1}^{n}Q_i^2 + \frac{K}{D^{4.87}}\left(\frac{Q_i}{C}\right)^{1.852} \times \frac{L_i}{100}.$$
(7.10)

Assuming the uniform lateral spacing, $L_s = (h_r + W_t)$, we get:

$$L_1 = L_2 = L_3 = \ldots = L_i = \ldots = L_n = L_s = (h_r + W_t)$$

$$h_{fb} = 2\frac{K_b K_r}{D^4}\sum_{i=1}^{n}Q_i^2 + \frac{K}{D^{4.87}} + \frac{K.L_s}{100.C^{1.852}.D^{4.87}}\sum_{i=1}^{n}Q_i^{1.852}.$$
(7.11)

Let $K_1 = 2 \cdot K_b \cdot K_r$; and $K_2 = \dfrac{K.L_s}{100.C^{1.852}}$, then Equation (7.11) may be rewritten as:

$$h_{fb} = \frac{K_1}{D^4}\sum_{i=1}^{n}Q_i^2 + \frac{K_2}{D^{4.87}}\sum_{i=1}^{n}Q_i^{1.852}.$$
(7.12)

The Equation (7.12) is the general equation for the calculation of head loss due to friction and pipe bends. Using this equation, the head loss from the first terrace to any number of terraces (in between first and last terrace) can be calculated. This equation can also be used for the calculation of total head loss, when the water source is at the top or bottom of any shape of terraced or nonterraced land. When water is available at the top of the terraced land or field, the design is based on the gravity head or zero energy condition for the micro irrigation system. After designing the main/sub-main, lateral lines are designed based on 10% permissible head loss. The emitters/drippers are selected according to the soil type and crop to be irrigated using general design procedure of micro irrigation system.

## 7.3   DESIGN CHARTS

The design charts for the design of main/sub-main (manifolds) for the ground slopes up to 100% was developed (Fig. 7.2). The design for the ground slopes of 50% and 75% are presented here for 10 and 5 numbers of terraces, respectively. These slopes cover almost all the hilly horticulture lands in India. The 25% slope was not considered in the development of design chart in India, because people commonly adapt the vegetable or

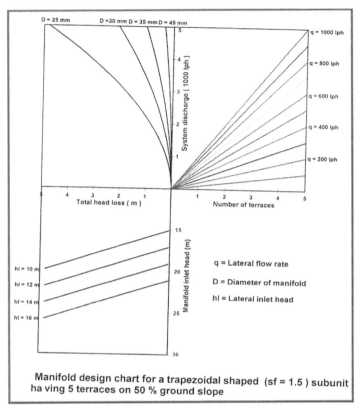

**Manifold design chart for a trapezoidal shaped (sf = 1.5 ) subunit ha ving 5 terraces on 50 % ground slope**

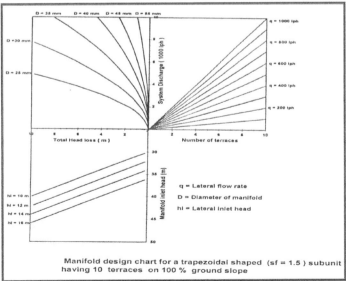

**Manifold design chart for a trapezoidal shaped (sf = 1.5 ) subunit having 10 terraces on 100 % ground slope**

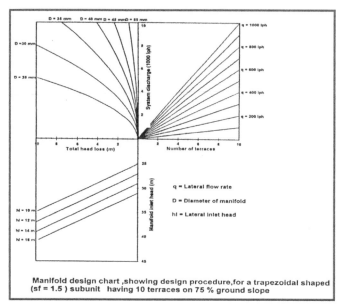

**FIGURE 7.2**    Design charts for manifold for 50%, 75%, and 100% slope of terraced land.

cereals farming even up to 35% slopes and generally do not take horticultural crops in such low-sloped lands.

Table 7.1 presents the head loss calculated from theoretical design procedure and head loss obtained using design charts. This table reveals that the percentage errors for the results obtained by two methods are well within 5%. This is acceptable for the micro irrigation design to achieve the uniformity of water application to the level of more than 90%.

**TABLE 7.1**    Head Loss Calculated from Theoretical Design Procedure and Design Charts.

| Shape factor ($S_f$) | Diameter of manifold (mm) | Percentage error in total head loss | | | | | |
|---|---|---|---|---|---|---|---|
| | | 60% ground slope | | | 80% ground slope | | |
| | | Extrapolated values (m) | Calculated values (m) | % error | Extrapolated values (m) | Calculated values (m) | % error |
| 0.5 | 30 | 4.96 | 4.888 | 1.45 | 5.44 | 5.229 | 4.04 |
| 0.5 | 35 | 2.5 | 2.427 | 2.92 | 2.66 | 2.588 | 2.78 |
| 0.5 | 40 | 1.34 | 1.324 | 1.19 | 1.42 | 1.408 | 0.85 |
| 0.5 | 45 | 0.77 | 0.776 | 0.78 | 0.82 | 0.823 | 0.36 |

**TABLE 7.1**  *(Continued)*

| Shape factor ($S_f$) | Diameter of manifold (mm) | Percentage error in total head loss | | | | | |
|---|---|---|---|---|---|---|---|
| | | 60% ground slope | | | 80% ground slope | | |
| | | Extrapolated values (m) | Calculated values (m) | % error | Extrapolated values (m) | Calculated values (m) | % error |
| 1.0 | 30 | 6.22 | 6.214 | 0.01 | 6.70 | 6.647 | 0.80 |
| 1.0 | 35 | 2.96 | 3.085 | 4.05 | 3.26 | 3.290 | 0.99 |
| 1.0 | 40 | 1.64 | 1.684 | 2.61 | 1.76 | 1.790 | 1.68 |
| 1.0 | 45 | – | 0.987 | – | 1.02 | 1.047 | 2.58 |
| 1.5 | 30 | 7.96 | 7.836 | 1.58 | 8.70 | 8.380 | 3.82 |
| 1.5 | 35 | 3.98 | 3.892 | 2.26 | 4.20 | 4.149 | 1.23 |
| 1.5 | 40 | 2.18 | 2.124 | 2.60 | 2.34 | 2.258 | 3.63 |
| 1.5 | 45 | 1.24 | 1.245 | 0.40 | 1.32 | 1.321 | 0.08 |

## 7.4  SUMMARY

The results in this chapter can be summarized as follows: (1) Gravity fed zero energy micro irrigation system can be easily designed using the developed theoretical procedure; (2) Field micro irrigation engineer can use design charts for the design of micro irrigation system in hilly terraced land under prevailing slope, water source, and soil conditions.

## KEYWORDS

- cash crops
- drippers
- emitter discharge
- friction coefficient
- Hazen–Williams equation

## REFERENCES

1. Keller, J.; D. Karmeli. Trickle Irrigation Design Parameters. *Trans. ASAE.* **1974,** *17*(4), 678–684.
2. Shukla, K. N.; Singh, P. K.; Singh, K. K.; Pandey, P. K. Water Requirement and Method of Irrigation for Hill Horticulture. *Souvenir, National Conference on Micro Irrigation (NCOM),* GBPUAT, Pantnagar, India, June 3–5, 2005; 64–70.

# CHAPTER 8

# UNIFORMITY MEASUREMENT METHODS: SPINNER-TYPE MICRO-SPRINKLERS

M. V. MANJUNATHA[1], SURJEET SINGH[2], and AJAI SINGH[3*]

[1]Department of Agricultural Engineering, University of Agricultural Sciences, Dharwad 580005, Karnataka, India

[2]National Institute of Hydrology, Roorkee, Uttarakhand, India

[3]Centre for Water Engineering and Management, Central University of Jharkhand, Ranchi, Jharkhand, India

*Corresponding author. E-mail: ajai_jpo@yahoo.com; ajai.singh@cuj.ac.in

## CONTENTS

## 8.1   INTRODUCTION

Agriculture draws more than 85% of the total water used in India. It is estimated that allocation of water to agriculture will reduce to 71% in the next 25 years, due to expected increase in demand of water for industries and municipal purposes. Technical innovations need to be exploited to achieve the twin objectives of higher productivity and optimum use of water. This calls for adoption of advanced methods of irrigation like drip, sprinkler, and micro-sprinkler for improved and efficient water management. The salient features of micro-sprinkler irrigation system are its higher water use efficiency and water savings compared with sprinkler irrigation system. It has lower operating pressure requirement ($1–2$ kg/cm$^2$) and low initial cost in comparison with drip irrigation especially for field crops. It is less susceptible to clogging and hence has low maintenance cost. It has more area of coverage than drip emitters with a single micro-sprinkler and is suitable for frost protection, cooling of greenhouses, poultry houses, nurseries, and under tree irrigation. It facilitates maximum output per unit of water because of lesser evaporation losses than sprinkler system, and not only has lesser energy requirement than it but also more ease in handling. An optimum sprinkler design requires that water should be applied to the crop in a uniform manner, which mainly depends on the type of sprinkler, nozzle size, operating pressure, and sprinkler spacing. Kerr et al. [7] reported that the acceptable Christiansen's uniformity coefficient (UCC) could be achieved only after 75% × 75% overlaps of the sprinklers. Nimah et al. [8] reported that distribution uniformity was improved with increasing wind speeds and operating pressure but pattern application efficiency was affected by evaporation and wind drift losses.

Keeping these considerations in view, studies were conducted at Pantnagar, India, to evaluate the performance of various micro-sprinklers under different operating pressures.

## 8.2   METHODS TO MEASURE THE UNIFORMITY COEFFICIENT OF MICRO-SPRINKLERS

### 8.2.1   CHRISTIANSEN'S UNIFORMITY COEFFICIENT (UCC)

Catch-can measurements are used to determine the uniformity of a sprinkler irrigation system. Christiansen [4] was the first to develop a numerical index representing the system uniformity of overlapping sprinklers. His uniformity coefficient (UCC) is a percentage on a scale of 0–100 (absolute uniformity). A uniformity coefficient of 70% is considered by many investigators to be the minimum acceptable performance. Higher uniformity coefficients are usually needed with intensively maintained ornamentals. Catch-can measurements are also used to illustrate water distribution or patterns. UCC is defined as:

$$UCC = 100 \, [1 - (x/mn)], \tag{8.1}$$

where UCC is the Christiansen's uniformity coefficient; x is the sum of the absolute deviations of individual observations $(x_i)$ from the mean of the observations (m); and n is the number of observations. All deviations from the mean are positive numbers. Therefore, any negative number is changed to a positive number. For example, given a mean of 35 ml and observation of 31 ml would have a deviation of 4 (31−35 = −4 = 4).

## 8.2.2   HART UNIFORMITY COEFFICIENT (UCH) AND PATTERN EFFICIENCY (PE_H)

Hart [5] developed the uniformity coefficient in association with *Hawaiian Sugar Planter's Association*. His uniformity coefficient is based on normal (Gaussian) distribution. If water application depths are assumed to be normally distributed, the UCC and UCH will have the same value. The UCH is defined as:

$$UCH = \{1 - [0.798 \, (s/m)]\} \tag{8.2}$$

$$PE_H = \{1 - (1.27 \, s/m)\}, \tag{8.3}$$

where UCH is the Hart uniformity coefficient [5], $PE_H$ is the Hart pattern efficiency [5], s is the standard deviation, and m is the mean of the observations.

### 8.2.3  CHOWDHARY UNIFORMITY COEFFICIENT ($U_C$) AND PATTERN EFFICIENCY ($PE_C$)

Chowdhary [3] took into consideration the skewness of the precipitation distribution to develop his uniformity coefficient. It is defined as:

$$U_C = \{1 - [(0.798 \text{ s/m})] [1/(1 + C_s^2/48)]\} \tag{8.4}$$

$$PE_C = \{1 - [1.27 / (1+0.175 \text{ } C_s^{1.5}) \text{ (s/m)}]\}, \tag{8.5}$$

where $U_C$ is the Chowdhary uniformity coefficient; $PE_C$ is the Chowdhary pattern efficiency; s is the standard deviation; m is the mean of the observations; and $C_s$ is the coefficient of skew.

## 8.3  MATERIALS AND METHODS

The experiment was conducted at G. B. Pant University of Agriculture and Technology, Pantnagar (29°N latitude, 79°18′E longitude, 243.9 m above MSL), during the months of September and October. The tests were conducted on a cement concrete floor having a slope of approximately 1%, which is within the permissible limits (2%) as recommended by ASAE [1] and also by B1S [6]. The average temperature, humidity, wind speed, and evaporation during the test period were 24.7°C, 76.5%, 1.94 km/h, and 3.35 mm, respectively.

### 8.3.1  EXPERIMENTAL LAYOUT AND PROCEDURE

Due to lack of testing standards for micro-sprinklers, the American Society of Agricultural Engineers Standard, ASAE S330.1, "Procedure for sprinkler distribution testing for research purpose [1]" was used as a guideline for this study. The variables selected for the study were: three operating pressures (1.0, 1.5, and 2.0 kg/cm²) and eight different micro-sprinklers (make: A, B, C, D, E, F, G, and H). Since the stakes provided by the manufacturers were of 35 cm, the riser height was kept constant as 35 cm. Catch cans were used to collect water from micro-sprinklers at a grid spacing of 0.3 m × 0.3 m in one-quarter of 6.5 m × 6.5 m covering area of micro-sprinklers as suggested by Boman [2]. The micro-sprinklers were run for a period of 1 h, so that the amount of water catched in each catch can was easily measurable. The volume of water collected in the catch cans was thus converted into precipitation rate (mm/h) for further use.

Utilizing the depth distribution data of a single nozzle, overlapping patterns of various micro-sprinkler spacings (3 m × 3 m–10 m × 10 m) were obtained. Different uniformity coefficients developed by Christiansen [4], Hart [5], and Chowdhary [3], and pattern efficiencies of Hart [5] and Chowdhary [3] for sprinklers were computed and compared for each of the different overlapping patterns.

## 8.4 RESULTS AND DISCUSSION

### 8.4.1 UNIFORMITY COEFFICIENTS

In general, UCC was increased with increase in operating pressure (from 1 to 2 kg/cm$^2$) and was decreased with increase in micro-sprinkler spacings (3 m × 3 m–10 m × 10 m) considered in this study (Table 8.1). The decrease in UCC appears to be sharper at lower operating pressures than at higher operating pressures for most of the micro-sprinklers. At a micro-sprinkler spacing of 3 m × 3 m and an operating pressure of 1.0, 1.5, and 2.0 kg/cm$^2$, the highest UCC of 91%, 92%, and 96% were observed for make H, H, and C micro-sprinklers, respectively. It was also found that even at micro-sprinkler spacing of 10 m × 10 m and at an operating pressure of 2 kg/cm$^2$, about 70% of UCC was observed in case of make H micro-sprinkler, which showed its better performance compared with other micro-sprinklers. The desired UCC of 70% was obtained for all micro-sprinklers at closer spacings of 3 m × 3 m and at operating pressures of 1.0 and 1.5 kg/cm$^2$. More than 90% UCC was achieved at a spacing of 3 m × 3 m at an operating pressure of 2.0 kg/cm$^2$. To maintain the desired UCC of greater than 70%, the optimum micro-sprinkler spacings were 7 m × 7 m, 5 m × 5 m, 8 m × 8 m, 4 m × 4 m, 6 m × 6 m, 7 m × 7 m, 4 m × 4 m, and 10 m × 10 m for make A, B, C, D, E, F, G, and H micro-sprinklers, respectively, at an operating pressure of 2.0 kg/cm$^2$.

In order to relate the effect of different operating pressures and micro-sprinkler spacings with UCC, multiple linear regression analysis was conducted for all eight micro-sprinklers (Table 8.2)

**TABLE 8.1**    Christiansen's Uniformity Coefficient (UCC) for Various Micro-sprinklers.

| Operating pressure (kg/cm²) | Micro-sprinkler spacing (m) | Christiansen's uniformity coefficient, UCC (%) | | | | | | | |
|---|---|---|---|---|---|---|---|---|---|
| | | Makes of micro-sprinkler | | | | | | | |
| | | A | B | C | D | E | F | G | H |
| 1.0 | 3 × 3 | 79 | 78 | 77 | 75 | 76 | 89 | 86 | 91 |
| | 4 × 4 | 56 | 75 | 69 | 66 | 71 | 76 | 60 | 89 |
| | 5 × 5 | 26 | 61 | 60 | 48 | 52 | 75 | 37 | 73 |
| | 6 × 6 | 22 | 42 | 54 | 44 | 37 | 69 | 35 | 61 |
| | 7 × 7 | 20 | 30 | 48 | 41 | 9 | 54 | 34 | 61 |
| | 8 × 8 | 15 | 17 | 34 | 23 | – | 37 | 16 | 59 |
| | 9 × 9 | – | 7 | 12 | 8 | – | 19 | 2 | 58 |
| | 10 × 10 | – | – | – | – | – | 3 | – | 47 |
| 1.5 | 3 × 3 | 75 | 85 | 80 | 83 | 78 | 91 | 88 | 92 |
| | 4 × 4 | 65 | 76 | 82 | 61 | 75 | 82 | 58 | 91 |
| | 5 × 5 | 56 | 64 | 69 | 51 | 69 | 74 | 39 | 90 |
| | 6 × 6 | 49 | 47 | 67 | 41 | 57 | 72 | 36 | 73 |
| | 7 × 7 | 29 | 35 | 54 | 40 | 34 | 63 | 27 | 71 |
| | 8 × 8 | 20 | 18 | 53 | 34 | 10 | 47 | 23 | 70 |
| | 9 × 9 | – | 10 | 37 | 24 | – | 28 | 13 | 68 |
| | 10 × 10 | – | – | 16 | 7 | – | 13 | – | 67 |
| 2.0 | 3 × 3 | 94 | 92 | 96 | 91 | 95 | 93 | 94 | 95 |
| | 4 × 4 | 87 | 80 | 82 | 82 | 83 | 83 | 80 | 93 |
| | 5 × 5 | 83 | 77 | 80 | 65 | 81 | 82 | 65 | 87 |
| | 6 × 6 | 77 | 61 | 79 | 63 | 79 | 79 | 55 | 76 |
| | 7 × 7 | 75 | 46 | 78 | 59 | 66 | 70 | 41 | 76 |
| | 8 × 8 | 64 | 25 | 77 | 57 | 61 | 59 | 37 | 75 |
| | 9 × 9 | 55 | 12 | 67 | 56 | 55 | 45 | 30 | 72 |
| | 10 × 10 | 35 | – | 42 | 34 | 37 | 32 | 15 | 70 |

**Note:** "–" indicates uniformity coefficient is zero or negative in all Tables 8.1–8.6.

$$y = a_1 + b_1 x_1 + b_2 x_2 \qquad (8.6)$$

where y = estimated Christiansen's uniformity coefficient (%); $x_1$ = operating pressure (kg/cm²); $x_2$ = micro-sprinkler spacing (m); $a_1$, $b_1$, and $b_2$ are empirical constants (regression coefficients). The fitted empirical equations are presented in Table 8.2.

**TABLE 8.2**  Multiple Regression Equations Relating UCC with Pressure and Spacings for Eight Micro-sprinklers.

| Makes of micro-sprinkler | Fitted equation | Correlation coefficient (r) |
|---|---|---|
| A | $y = 38.7678 + 44.919\, x_1 - 9.187\, x_2$ | 0.935 |
| B | $y = 111.000 + 11.857\, x_1 - 13.226\, x_2$ | 0.990 |
| C | $y = 67.040 + 28.634\, x_1 - 7.775\, x_2$ | 0.934 |
| D | $y = 67.682 + 24.530\, x_1 - 8.645\, x_2$ | 0.942 |
| E | $y = 59.525 + 37.095\, x_1 - 10.171\, x_2$ | 0.907 |
| F | $y = 104.844 + 15.125\, x_1 - 10.408\, x_2$ | 0.962 |
| G | $y = 82.051 + 20.389\, x_1 - 11.155\, x_2$ | 0.959 |
| H | $y = 85.010 + 13.125\, x_1 - 4.575\, x_2$ | 0.929 |

$y = a_1 + b_1 x_1 + b_2 x_2$, where: $y$ = calculated Christiansen's uniformity coefficient (%), $x_1$ = operating pressure (kg/cm$^2$), $x_2$ = micro-sprinkler spacing (m), and the regression coefficients are $a_1$, $b_1$, and $b_2$

The trend of variations of UCH and $U_C$ are similar to that of Christiansen's uniformity coefficient (UCC) for all eight micro-sprinklers at various operating pressures and spacings (Tables 8.3 and 8.4). Larger variation was observed at lower operating pressure compared with relatively higher operating pressure of 2 kg/cm$^2$ for micro-sprinklers A, C, D, E, and F. However, sharp declining trend of UCH and $U_C$ was observed for make B and G micro-sprinklers at all the operating pressures. This is mainly due to lesser area of coverage and nonuniform distribution of water within the overlapped area. However, in case of make H micro-sprinkler, a gradual reduction in uniformity with increase in sprinkler spacing (from 3 m to 10 m) was observed at all three levels of operating pressures. Highest (96%) uniformity was observed for make C and E micro-sprinklers at 3 m × 3 m spacing at an operating pressure of 2.0 kg/cm$^2$ and was lowest (3%) in case of make B micro-sprinkler at 10 m × 10 m spacing. For the same pressure (2 kg/cm$^2$), the uniformity was above 90% for all the makes of micro-sprinklers at 3 m × 3 m spacing. A fairly good uniformity coefficient (68%) was observed in case of make H micro-sprinkler at the operating pressure of 2.0 kg/cm$^2$ even at the wider spacing of 10 m × 10 m, thus indicating its better water distribution compared with other makes of micro-sprinklers. A better uniformity (>70%) was also recorded in case of make C micro-sprinkler followed by make A and D micro-sprinklers

(>60%) at the operating pressure of 2.0 kg/cm² and at the spacing of 8 m × 8 m.

**TABLE 8.3**   Hart Uniformity Coefficient (UCH) for Eight Micro-sprinklers.

| Operating pressure (kg/cm²) | Micro-sprinkler spacing (m) | Hart uniformity coefficient (%) | | | | | | | |
|---|---|---|---|---|---|---|---|---|---|
| | | Makes of micro-sprinkler | | | | | | | |
| | | A | B | C | D | E | F | G | H |
| 1.0 | 3 × 3 | 77 | 82 | 78 | 75 | 80 | 91 | 85 | 91 |
| | 4 × 4 | 56 | 76 | 64 | 52 | 68 | 78 | 51 | 89 |
| | 5 × 5 | 33 | 63 | 60 | 50 | 59 | 74 | 46 | 75 |
| | 6 × 6 | 30 | 44 | 56 | 49 | 39 | 68 | 43 | 62 |
| | 7 × 7 | 25 | 30 | 50 | 40 | 17 | 52 | 38 | 61 |
| | 8 × 8 | 15 | 14 | 39 | 29 | – | 35 | 23 | 60 |
| | 9 × 9 | – | – | 22 | 13 | – | 19 | 6 | 60 |
| | 10 × 10 | – | – | 6 | – | – | 3 | – | 47 |
| 1.5 | 3 × 3 | 77 | 86 | 79 | 82 | 80 | 91 | 87 | 93 |
| | 4 × 4 | 62 | 79 | 83 | 65 | 75 | 82 | 60 | 95 |
| | 5 × 5 | 61 | 68 | 68 | 55 | 66 | 76 | 47 | 91 |
| | 6 × 6 | 50 | 52 | 65 | 45 | 48 | 73 | 42 | 73 |
| | 7 × 7 | 31 | 38 | 56 | 40 | 29 | 62 | 26 | 70 |
| | 8 × 8 | 12 | 21 | 54 | 39 | 11 | 46 | 17 | 69 |
| | 9 × 9 | – | 9 | 40 | 31 | – | 30 | 11 | 69 |
| | 10 × 10 | – | – | 25 | 16 | – | 15 | – | 65 |
| 2.0 | 3 × 3 | 93 | 93 | 96 | 90 | 96 | 93 | 94 | 95 |
| | 4 × 4 | 86 | 80 | 83 | 82 | 85 | 85 | 77 | 93 |
| | 5 × 5 | 82 | 75 | 80 | 67 | 79 | 81 | 65 | 88 |
| | 6 × 6 | 80 | 64 | 79 | 63 | 75 | 80 | 51 | 77 |
| | 7 × 7 | 77 | 52 | 78 | 62 | 61 | 70 | 49 | 76 |
| | 8 × 8 | 63 | 31 | 73 | 61 | 50 | 57 | 32 | 73 |
| | 9 × 9 | 55 | 18 | 64 | 57 | 44 | 44 | 28 | 73 |
| | 10 × 10 | 38 | 3 | 46 | 40 | 32 | 30 | 13 | 68 |

**TABLE 8.4**  Chowdhary Uniformity Coefficient (UC) for Eight Micro-sprinklers.

| Operating pressure (kg/cm²) | Micro-sprinkler spacing (m) | Chowdhary uniformity coefficient (%) | | | | | | | |
|---|---|---|---|---|---|---|---|---|---|
| | | Makes of micro-sprinkler | | | | | | | |
| | | A | B | C | D | E | F | G | H |
| 1.0 | 3 × 3 | 77 | 82 | 78 | 75 | 80 | 91 | 85 | 91 |
| | 4 × 4 | 57 | 76 | 66 | 58 | 59 | 74 | 53 | 89 |
| | 5 × 5 | 34 | 63 | 56 | 52 | 68 | 78 | 46 | 75 |
| | 6 × 6 | 31 | 44 | 60 | 42 | 39 | 68 | 40 | 62 |
| | 7 × 7 | 26 | 30 | 51 | 49 | 18 | 53 | 43 | 61 |
| | 8 × 8 | 16 | 16 | 39 | 30 | – | 36 | 24 | 62 |
| | 9 × 9 | – | 7 | 22 | 14 | – | 21 | 8 | 60 |
| | 10 × 10 | – | – | 7 | – | – | 7 | – | 48 |
| 1.5 | 3 × 3 | 77 | 86 | 79 | 82 | 80 | 92 | 87 | 93 |
| | 4 × 4 | 63 | 79 | 83 | 65 | 75 | 82 | 60 | 94 |
| | 5 × 5 | 61 | 68 | 68 | 56 | 67 | 76 | 47 | 91 |
| | 6 × 6 | 50 | 53 | 65 | 41 | 50 | 74 | 21 | 73 |
| | 7 × 7 | 31 | 38 | 56 | 45 | 31 | 62 | 43 | 70 |
| | 8 × 8 | 14 | 23 | 54 | 40 | 15 | 46 | 27 | 70 |
| | 9 × 9 | – | 12 | 40 | 31 | – | 31 | 13 | 69 |
| | 10 × 10 | – | – | 25 | 16 | – | 17 | – | 65 |
| 2.0 | 3 × 3 | 94 | 93 | 96 | 90 | 96 | 93 | 94 | 95 |
| | 4 × 4 | 83 | 81 | 83 | 83 | 85 | 81 | 78 | 93 |
| | 5 × 5 | 78 | 76 | 78 | 67 | 79 | 85 | 65 | 88 |
| | 6 × 6 | 86 | 64 | 81 | 62 | 75 | 80 | 53 | 77 |
| | 7 × 7 | 80 | 52 | 79 | 61 | 62 | 71 | 49 | 73 |
| | 8 × 8 | 63 | 32 | 74 | 63 | 51 | 58 | 35 | 76 |
| | 9 × 9 | 55 | 19 | 65 | 57 | 45 | 45 | 30 | 73 |
| | 10 × 10 | 38 | 5 | 46 | 40 | 33 | 32 | 15 | 68 |

**Note:** "–" indicates uniformity coefficient is zero or negative.

## 8.4.2 PATTERN EFFICIENCY

The trend in variations of $PE_H$ and $PE_C$ is more or less similar to that of uniformity coefficients for all eight micro-sprinklers at various operating pressures and spacings (Tables 8.5 and 8.6). In general, pattern efficiency increased with increase in operating pressure (from 1 to 2 kg/cm$^2$) and decreased with increase in micro-sprinkler spacings (from 3 m × 3 m to 10 m × 10 m). Highest (94%) efficiency was observed for make C micro-sprinkler at 3 m × 3 m spacing at an operating pressure of 2.0 kg/cm$^2$. Fairly good efficiency (>80%) was observed for all the makes of micro-sprinklers at 3 m × 3 m spacing. It was found that even at wider spacing (10 m ×10 m) about 50% of pattern efficiency was achieved only in case of make H micro-sprinklers at an operating pressure of 2 kg/cm$^2$. Performance of make C micro-sprinkler was also found better (57%) up to 8 m × 8 m spacing in both the methods and indicated their better performance compared with other makes of micro-sprinklers.

**TABLE 8.5** Hart Pattern Efficiency ($PE_H$) for Various Micro-sprinklers.

| Operating pressure (kg/cm²) | Micro-sprinkler spacing (m) | Hart pattern efficiency (%) | | | | | | | |
|---|---|---|---|---|---|---|---|---|---|
| | | Makes of micro-sprinkler | | | | | | | |
| | | A | B | C | D | E | F | G | H |
| 1.0 | 3 × 3 | 64 | 71 | 64 | 60 | 69 | 85 | 76 | 86 |
| | 4 × 4 | 31 | 63 | 42 | 21 | 34 | 58 | 23 | 82 |
| | 5 × | – | 41 | 30 | 24 | 49 | 64 | 14 | 61 |
| | 6 × 6 | – | 11 | 20 | 5 | 3 | 49 | 2 | 39 |
| | 7 × 7 | – | – | 36 | 19 | – | 24 | 9 | 36 |
| | 8 × 8 | – | – | 3 | – | – | – | – | 39 |
| | 9 × 9 | – | – | – | – | – | – | – | 36 |
| | 10 × 10 | – | – | – | – | – | – | – | 17 |
| 1.5 | 3 × 3 | 63 | 77 | 73 | 72 | 68 | 87 | 80 | 89 |
| | 4 × 4 | 40 | 67 | 66 | 44 | 61 | 72 | 36 | 92 |
| | 5 × 5 | 38 | 49 | 50 | 29 | 46 | 62 | 16 | 85 |
| | 6 × 6 | 21 | 24 | 30 | 4 | 18 | 58 | 9 | 58 |
| | 7 × 7 | – | 1 | 44 | 13 | – | 40 | – | 53 |
| | 8 × 8 | – | – | 27 | 5 | – | 14 | – | 51 |
| | 9 × 9 | – | – | 5 | – | – | – | – | 51 |
| | 10 × 10 | – | – | – | – | – | – | – | 44 |

**TABLE 8.5** *(Continued)*

| Operating pressure (kg/cm²) | Micro-sprinkler spacing (m) | Hart pattern efficiency (%) | | | | | | | |
| | | Makes of micro-sprinkler | | | | | | | |
| | | A | B | C | D | E | F | G | H |
| 2.0 | 3 × 3 | 90 | 90 | 93 | 84 | 93 | 89 | 90 | 92 |
| | 4 × 4 | 78 | 69 | 73 | 72 | 76 | 76 | 63 | 89 |
| | 5 × 5 | 72 | 61 | 65 | 48 | 66 | 68 | 44 | 81 |
| | 6 × 6 | 68 | 43 | 69 | 40 | 61 | 67 | 22 | 64 |
| | 7 × 7 | 64 | 24 | 66 | 38 | 39 | 53 | 19 | 57 |
| | 8 × 8 | 41 | – | 57 | 41 | 20 | 32 | – | 61 |
| | 9 × 9 | 28 | – | 43 | 32 | 11 | 11 | – | 58 |
| | 10 × 10 | 2 | – | 14 | 5 | – | – | – | 50 |

**TABLE 8.6** Chowdhary Pattern Efficiency ($PE_C$) for Various Micro-sprinklers.

| Operating pressure (kg/cm²) | Micro-sprinkler spacing (m) | Chowdhary pattern efficiency (%) | | | | | | | |
| | | Makes of micro-sprinkler | | | | | | | |
| | | A | B | C | D | E | F | G | H |
| 1.0 | 3 × 3 | 65 | 71 | 67 | 64 | 68 | 85 | 77 | 86 |
| | 4 × 4 | 38 | 63 | 60 | 58 | 53 | 63 | 37 | 82 |
| | 5 × 5 | – | 41 | 30 | 24 | 34 | 68 | 15 | 61 |
| | 6 × 6 | – | 14 | 37 | 20 | 4 | 51 | 19 | 40 |
| | 7 × 7 | – | – | 36 | 23 | – | 30 | 14 | 44 |
| | 8 × 8 | – | – | 3 | – | – | 9 | – | 45 |
| | 9 × 9 | – | – | – | – | – | – | – | 37 |
| | 10 × 10 | – | – | – | – | – | – | – | 18 |
| 1.5 | 3 × 3 | 63 | 77 | 73 | 72 | 68 | 87 | 80 | 89 |
| | 4 × 4 | 49 | 67 | 66 | 44 | 61 | 72 | 39 | 92 |
| | 5 × 5 | 38 | 49 | 50 | 32 | 57 | 64 | 16 | 85 |
| | 6 × 6 | 23 | 26 | 38 | 19 | 32 | 61 | 3 | 58 |
| | 7 × 7 | – | 9 | 52 | 17 | 8 | 46 | 18 | 53 |
| | 8 × 8 | – | – | 27 | 6 | – | 24 | – | 57 |
| | 9 × 9 | – | – | 5 | – | – | 5 | – | 56 |
| | 10 × 10 | – | – | – | – | – | – | – | 45 |

**TABLE 8.6**    *(Continued)*

| Operating pressure (kg/cm²) | Micro-sprinkler spacing (m) | Chowdhary pattern efficiency (%) | | | | | | | |
|---|---|---|---|---|---|---|---|---|---|
| | | Makes of micro-sprinkler | | | | | | | |
| | | A | B | C | D | E | F | G | H |
| 2.0 | 3 × 3 | 90 | 90 | 94 | 84 | 93 | 90 | 91 | 92 |
| | 4 × 4 | 79 | 70 | 76 | 75 | 77 | 76 | 70 | 89 |
| | 5 × 5 | 73 | 65 | 69 | 51 | 70 | 76 | 45 | 81 |
| | 6 × 6 | 70 | 45 | 69 | 40 | 62 | 71 | 42 | 64 |
| | 7 × 7 | 68 | 26 | 66 | 39 | 45 | 61 | 22 | 57 |
| | 8 × 8 | 41 | – | 57 | 41 | 34 | 43 | 18 | 62 |
| | 9 × 9 | 28 | – | 43 | 32 | 21 | 27 | 4 | 59 |
| | 10 × 10 | 2 | – | 14 | 5 | 4 | 11 | – | 50 |

## 8.5   CONCLUSIONS

Based on the micro-sprinkler tests, it was observed that highest UCC of 96% was recorded for make C micro-sprinkler at the operating pressure of 2.0 kg/cm². UCC was increased with increase in operating pressure and was decreased with increase in micro-sprinkler spacing. The trend of variations of UCH and $U_C$ are similar to that of UCC for eight micro-sprinklers at various operating pressures and spacings. Larger variation in UCH and $U_C$ was observed for make A, C, D, E, and F micro-sprinklers at the lower operating pressures (1.0 and 1.5 kg/cm²) compared to variation at 2.0 kg/cm² pressure. Greater than 90% uniformity and pattern efficiency can be achieved at a closer spacing of 3 m × 3 m at an operating pressure of 2.0 kg/cm². A better uniformity and pattern efficiency was recorded in case of make H micro-sprinkler even at the wider spacing of 10 m × 10 m. Based on the results of UCC, the performance of make H micro-sprinkler was found to be better than other makes of micro-sprinklers tested.

## 8.6   SUMMARY

The application and selection of the micro-sprinklers are usually made on the basis of uniformity of water distribution within the overlapped patterns, which depend on type of micro-sprinkler, nozzle size, operating pressure, and micro-sprinkler spacing. Eight makes of commercially available spinner type micro-sprinklers (A, B, C, D, E, F, G, and H of different nozzle sizes 1.50, 1.60, 1.73, 1.92, 2.00, 2.10, 2.34, and 2.80 mm, respectively) were used in order to evaluate the uniformity and distribution efficiency. Tests were conducted to measure the water depth and distribution patterns from a single leg test. Tests were conducted for three different operating pressures of 1.0, 1.5, and 2.0 kg/cm$^2$ at a riser height of 35 cm. Utilizing the depth distribution data of a single nozzle, overlapped patterns of various micro-sprinkler spacings (from 3 m × 3m to 10 m × 10 m) were obtained. Different uniformity coefficients developed by Christiansen [4], Hart [5], and Chowdhary [3], and pattern efficiencies of Hart [5] and Chowdhary [3] were estimated and compared for each of the different overlapping patterns.

The trend of variation of pattern efficiencies developed by Hart (PE$_H$) and Chowdhary (PE$_C$) was similar to that of UCC, UCH, and Uc for all the micro-sprinklers at various operating pressures and spacings. PE$_H$ and PE$_C$ varied significantly from one make to the other under the various operating conditions.

In general, uniformity of water distribution increases with increase in operating pressure for all the micro-sprinklers and decreases with increase in micro-sprinkler spacing. Based on the results of different uniformities, the performance of make H micro-sprinkler was better than other makes of micro-sprinklers tested.

## KEYWORDS

- micro-sprinkler
- micro-sprinkler spacing
- Chowdhary uniformity coefficient
- Hart pattern efficiency
- wider spacing

## REFERENCES

1. ASAE. *ASAE Standard: Procedure for Sprinkler Distribution Testing for Research Purpose*. S330.1. ASAE: St. Joseph, MI, USA, 1996.
2. Boman, B. J. Distribution Pattern of Micro Irrigation Spinner and Spray Emitters. *Appl. Eng. Agric..* **1989,** *5*(1), 50–56.
3. Chowdhary, F. H. Sprinkler Uniformity Measures and Skewness. *J. Am. Soc. Civil Eng..* **1976,** *102*(IR4), 425–433.
4. Christiansen, J. E. The Uniformity of Application of Water by Sprinkler Systems. *Agric. Eng.* **1941,** *22*(3), 89–92.
5. Hart, W. E. Overhead Irrigation Pattern Parameters. *Agric. Eng.* **1961,** *42*(7), 354–355.
6. IS 10802-1984. *Methods for Determination of Radius of Throw of Agricultural Sprinklers.* p 7.
7. Kerr, G. L.; Pochop, L. O.; Borrelli, J.; Anderson, D. A. Distribution Patterns of Home Lawn Sprinklers. *Trans. ASAE.* **1980,** *23*(2), 387–392.
8. Nimah, M. N.; Bashour, I.; Hamra, A. *Field Evaluation of Low Pressure Center Pivots.* ASAE Paper No. 85-2061. 1985

# INDEX

T - #0825 - 101024 - C328 - 229/152/15 - PB - 9781774636640 - Gloss Lamination